WHY WE FEEL

WHY
WE FEEL

The Science of Human Emotions

VICTOR S. JOHNSTON

HELIX BOOKS

PERSEUS BOOKS
Reading, Massachusetts

ISBN 0-7382-0109-X

Library of Congress Catalog Card Number: 98-89426

Perseus Books is a member of the Perseus Books Group

Jacket design by Sara Eisenman
Text design by Karen Savary
Set in 11.5-point Bembo by Pagesetters, Inc.

1 2 3 4 5 6 7 8 9—0302010099
First printing, March 1999

Find Helix Books on the World Wide Web at
http://www.aw.com/gb/

CONTENTS

PREFACE

∞

THE HUMAN BRAIN IS THE MOST INTRICATE AND complex entity on our planet and perhaps the most complicated device in the entire universe. There is absolutely nothing that we can see, think, feel, or know that does not depend on the workings of this tangled web of nerve cells. It is involved in every aspect of our daily life; it defines who we are and it dictates the way we view the world around us. But few of us realize that our brain creates what is, in effect, a virtual reality. This idea is so contrary to our common sense that few rational people would consider such a proposition to be an important scientific insight that is essential for understanding human nature, but it is.

We believe that we live in a world that is full of sounds and colors and smells and tastes because this is what we experience every day of our lives and there appears to be no reason for thinking otherwise. In *Why We Feel* I propose that we must abandon this common-sense view of reality and eventually accept the fact that our conscious experiences depend on the nature of our evolved neural processes and not on the nature of the events in the world that activate those processes. That is, although the external environment is teeming with electromagnetic radiation and air pressure waves, without consciousness it is both totally black and utterly silent.

Conscious experiences, such as our sensations and feelings, are nothing more than evolved illusions generated within biological brains.

The human brain did not evolve to accurately represent the world around us; it evolved only to enhance the survival of our genes. This Darwinian insight has profound implications for understanding every aspect of human nature. Many simple questions acquire a whole new meaning when viewed from this novel perspective. Do rotten eggs really smell bad? Are some people beautiful? Is it simply a fortunate accident that waste products, like our feces or urine, smell and taste unpleasant? Is tissue damage really painful? In *Why We Feel* I argue that such sensory feelings are not properties of molecules or events in the external world; they are the evolved adaptive illusions of a conscious mind.

Similar questions can be asked of more complex emotions. Do men and women experience the same feelings? Is premenstrual depression related to reproductive failure? Why do orgasms feel so good? Why do we have so many different feelings, like love, pride, fear, and sadness? All such feelings are the evolved "omens" of our reproductive success that amplify the consequences of physical and social events that have, or had, some bearing on our gene survival. I explore how this relationship came about, how it is maintained throughout our lives, and the important consequences of this arrangement for how we learn, reason, and understand the world around us.

Why We Feel is an adventure story about the evolution of human feelings, and how they have become woven into the fabric of our brain. It is a story about consciousness, emotions, free will, learning, memory, reasoning, ethics, aesthetics, and the enormous creative potential that has been bestowed upon us by our ancestral history. I hope that you enjoy the journey.

This book would not have been possible without the dedicated research of many students, the encouragement of many friends, and the valuable feedback and support of many colleagues. Throughout this book I have made referenced the work of many talented students: Mary Burleson, Craig Caldwell, Melissa Franklin, Gayle Luchini, Valerie Madrid, David Miller, Juan C. Oliver-Rodriguez, Paul Vonnahme, and Xiao Tian Wang. I thank all of them for their invaluable contributions. I am deeply grateful to my colleagues David Buss, Gary Cziko, Martin Daly, Randy Thornhill, David Trafimow, and Xiao Tian Wang for reviewing chapters or earlier versions of the entire manuscript. Their comments were highly constructive and the book has benefited greatly from their insights. I am also greatly indebted to Nina Andersen, John Cone, Lou Corl, Donald Giddon, J. D. and Myriam Jarvis, and Derek Partridge for their continuing support, friendship, and assistance with different aspects of this book.

As a faculty member in the Psychology Department at NMSU, I have enjoyed the support of a truly remarkable group of scientists who have encouraged and supported my research even when it was contrary to their own beliefs. I could never have written this book without them. I am also indebted to Amanda Bichsel Cook of Helix Books, for her patience and thoughtful editing of the manuscript.

Finally, I owe a very special thanks to my wife, Brandi, who has continually inspired me and given meaning to all of my endeavors. It is with love and gratitude that I dedicate this book to her and my wonderful children, Wendy, Justin, and Mantissa.

ONE

The Grand Illusion

WHEN I FIRST MET DAVID, HE WAS SITTING NER-vously on a small wooden chair in a psychiatrist's office. As a psychology student, I spent my summer vacations cleaning the floors of mental hospitals, and this was my very first opportunity to see a real clinical interview. The fourteen-year-old boy was disheveled and fragile, and he never raised his eyes above the level of the large wooden desk that separated him from the discerning eyes of the attentive psychiatrist. "And where are the monsters now?" asked the doctor. The little red-haired boy remained silent, apparently lost in his own thoughts. Then suddenly, as if awakening from a dream, David stood, turned around, and pointed toward the door. "They're in there," he said, probing at a very small black dot that I am sure no one, before that day, had ever noticed. "They're just waiting for you to go."

After the interview I sat patiently as the psychiatrist filled out a voluminous pile of forms, but at the first opportunity I asked, "What do you think?" The reply disturbed me. "Schizophrenia," he asserted. "Early stages. But we'll admit him to the hospital today." "What will happen to him?" I asked, feeling a wave of sadness spread over my body. "Probably spend the rest of his life here," he replied as he checked off more boxes on the papers in front of him. "Lost reality contact!" he continued, as if this would make everything clear to me. But nothing was clear. I wondered what was going on inside David's head. What was it like to experience a different reality?

Plainly David had great difficulty getting along in the world in ways that you and I take for granted. For David, the lurking monsters were his reality. He was as convinced of their existence as I was convinced that they didn't exist. For a moment I wondered what it would be like if the psychiatrist also saw the monsters and I were all alone in my personal reality; the thought frightened me. I had a glimpse of David's condition and could feel his fear. There was no doubt that his mind was disturbed—in fact, some might say that he had lost his mind—but the problem was manifested by his loss of contact with reality. He didn't see or interact with the world in an appropriate way, like the rest of us. It was many years before I gained some insight into the tenuous links between mind and reality, how they are established, and how easily they can be shattered. I also began to realize that even the mentally "well" among us have a distorted picture of the physical world and that this misperception has prevented us from understanding what may be the essence of our humanity: our emotions. Indeed, we are not as different from David as some of us might like to think.

Most of us believe that there is a real world "out there" that we can experience through our senses; we can see it, hear it, touch it, smell it, and taste it. Furthermore, we believe our picture of the world is, for the most part, an accurate reflection of that reality. That is, we believe that a sweet red apple appears red because it really is red, and that it tastes sweet because sweetness is an inherent property of the sugars inside apples. Of course, we also know that sometimes our senses can be deceived. We have all experienced visual illusions. For example, if we stare at a red apple for a few minutes and then quickly look at a white wall, we perceive what appears to be a green apple on the wall. This mental image is clearly an illusion since the apple isn't really there and it's the wrong color. Our prolonged staring at the red apple has depleted the "red" pigment in the visual receptors of a small area of our retina, and the reduced adigm, however, selection acts on the emergent properties, and the actual physical design of future cars will be a consequence of the successful, functional emergent properties that allowed their predecessors to complete the course. After several generations it is apparent that although the physical components do all the work, their arrangement—how they get to be organized the way they are—is a consequence of the emergent properties arising from that arrangement. Analogously, nerve cells are certainly the active agents in the nervous system, but their organization depends on the survival value of the emergent properties that arise from that organization. Over the long course of evolution, the functional attributes of the mind have been responsible for shaping the physical and chemical structure of brains. From this viewpoint, functional consequences dictate structural design.

For a biological example, consider the case of the peppered moths (*Biston betularia*) in Birmingham, England, at the time of the Industrial Revolution. Before the city and its surroundings became polluted by industrial waste, the moths were predominately white. As the environment became blackened, birds easily detected the white moths. As white moth numbers dwindled, a black moth population exploded. Note that natural selection did not favor either white or black moths; rather, the selection process favored the emergent property of "camouflage." Whiteness and blackness have no inherent value in and of themselves; it is "camouflage" that has survival value. Thus physical structures are merely secondary consequences, of their functional consequences, and functional attributes direct structural design. An enzyme is simply a protein unless it plays a functional role. But if a protein has a functional role—in digestion, for example—then it has survival value, and any gene that codes for this enzyme will increase in future generations.

The hard problem of consciousness has now been partially resolved but the specific functions of the emergent properties of mind still require some elucidation.

Not every emergent property is functional; "noisiness" played no role in our imaginary car race. In biological organisms only the functional attributes that ensure survival and reproduction will eventually be transmitted to future generations. From this perspective, the attributes of mind are not just any emergent properties of the neural organization; they are those functional emergent properties that enhance biological survival. That is, of all the emergent properties that could emanate from neural networks, only a subset will be selected: the subset that ensures the survival of DNA that can design brains that possess these emergent properties. Rather than creating a general-purpose computer, this iterative design process ensures that the evolved attributes of mind have a very special function—gene survival. Nor are these attributes independent of their biological underpinnings, as DCS suggests; rather they are selected because they help to preserve and perpetuate the biological tissue to which they owe their very existence. Relentless selection slowly but inexorably leads to adaptive or at least "satisficing"[4] functional design. This simple premise has profound implications for understanding the nature and origin of all conscious experiences. More specifically, the design and functional significance of human feelings now become apparent.

If rotten eggs smell bad, tissue damage causes pain, or sugar tastes sweet, it is not because hydrogen sulfide gas has a foul smell, or because pain is waiting to escape from the point of a needle as it enters the skin, or because sweetness is a property of sugar molecules. Rather, it is because the human brain has evolved a neural organization that can generate pleasant or unpleasant sensations for those aspects of the world that are a benefit or detriment to gene survival. That is, only organisms that have evolved such evaluative subjective feelings have been successful in transmitting their genes to successive generations. The individual organism need not be aware of the relationship between a foul odor and bacterial contamination, between tissue damage and infection, or between a sweet taste and the manufacture of ATP (the energy molecule in the body); natural selection has already established this relationship between emergent conscious feelings and gene survival. Most of us are oblivious to the fact that the energy required to contract our muscles, transport substances across our cell membranes, and make the important chemicals required by every living cell of our body is supplied by breaking one of the high-energy phosphate bonds of ATP to produce ADP.[5] However, we don't need to know or understand these mechanisms in order to survive; that knowledge is already part of our biological nature.

Sugars, a very rich source of dietary energy, simply taste good. Sweetness, however, is not a property of a sugar molecule; it is an evolved emergent property of our brain. Such sensations provide us with an immediate evaluation of sensory events, even in the absence of any understanding of their biological importance or their evolutionary origins. Our discomfort at high or low temperatures and the unpleasant smell of our waste products are both evolved emergent properties that were elicited by environmental events that consistently posed a threat to biological survival in ancestral environments. Individuals require no knowledge of these relationships, because natural selection has already forged the link between our conscious feelings and gene survival. We are aware of our *proximate design,* our sensations, and our feelings, but we have little awareness of how these emergent properties are related to the survival of our genes, our *ultimate design.*

Normally, we don't ask why we don't see air pressure waves or hear electromagnetic radiation, or why we don't generate a completely different subjective experience (which we can't imagine) in the presence of gamma radiation or when we are traveling at a constant velocity. Under those circumstances we would perceive our world quite differently. The way we do see it, therefore, requires some justification. But the illusion of naive realism is so powerful and ubiquitous that we come to believe that objects really are red, or hot, or bitter, or sweet, or beautiful, and we usually do not ask how or why we impose this structure on our physical world, or how this structure relates to our biological survival. We talk about the world around us as if it is full of light and sounds and tastes and smells. The physical world certainly contains electromagnetic radiation, air pressure waves, and chemicals dissolved in air or water, but not a single light or sound or smell or taste exists without the emergent properties of a conscious brain. Our conscious world is a grand illusion!

INTERACTING WITH THE PHYSICAL WORLD

If we have learned anything from twentieth-century physics, it is the realization that the physical world is composed of energy, some of which is currently in a form that we call *matter*. A subjective experience, like "redness," does not exist in this external world; rather, the brain has evolved to generate this particular experience in the presence of a specific frequency of electromagnetic radiation. A completely different experience, "greenness," is produced in response to a very similar frequency. Indeed, the physical difference in wavelength between "red" and "green" is a mere 150 billionths of a meter; they are essentially identical. In contrast, the room that you are in is filled with the infrared (IR) emissions from your body, plus signals from the local radio station (AM). Although the AM signals are more than a billion times the wavelength of the IR radiation, neither signal is detected, and no discrimination is made between them. Our senses and subjective experiences seem to have evolved to generate a nonlinear[6] representation of the energy/matter that exists "out there." While some specific energy/matter configurations elicit vivid subjective experiences, others are completely ignored. What biological benefits could have led to this arrangement?

For most animals, visual perception is limited to detecting and discriminating within a small range of electromagnetic frequencies centered around those reflected by the leaves of plants. This is not surprising since the survival of most animals depends on a food chain based on the ability of plants to convert solar energy into sugars using photosynthesis. The middle of our visible spectrum, which we perceive as green, corresponds to the frequencies reflected by the chlorophyll molecule when exposed to the incident radiation from our sun, or so-called white light. However, the incident radiation from the sun varies with weather conditions, so to compensate for these variations, it becomes necessary for animals to be able to detect other frequencies. Direct sunshine contains a wide range of frequencies, but radiation reflected from the sky is mainly in the higher frequency range (the sky appears blue) while lower frequencies dominate when the sun's radiation is scattered by particulate matter (the sky often appears red at sunrise or sunset). Under these different environmental conditions, the frequencies reflected by a "green" leaf vary considerably. If our perception of a leaf's color were simply a function of the reflected radiation, then the leaf would change color at sunset or when a cloud moved in front of the sun, causing it to be illuminated by radiation reflected from the sky. Remarkably, this doesn't happen. We possess a mechanism for color constancy that can compensate for these variations in incident radiation. To perceive a leaf as green under all the different conditions on earth, it is necessary to detect and compensate for the range of incident radiation that commonly bathes our planet. That is, in addition to "green" detectors, we require both "red" and "blue" detectors. The cones of our retina, and our conscious experiences of color, seem to have evolved in response to the very specific and perhaps unique problems encountered by diurnal animals[7] that depend on a chlorophyll-based food chain on a planet that possesses particular atmospheric conditions.[8]

If an organism experiences one emergent property (redness) when exposed to one frequency of electromagnetic radiation and a completely different emergent property (greenness) when exposed to a frequency that is physically almost identical, then it is not only discriminating between these signals but also exaggerating the difference between them. Such an organism enjoys the functional benefits of this discrimination, so the retinal and neural organization underlying these emergent properties will continue to be refined over generations. Similarly, an organism that experiences a discrete conscious feeling (sweetness) when presented with a valuable resource (sugars) and a very different feeling (sourness) when faced with a common waste product (acids) will pass on to future generations the neural circuitry that underlies this ability to form discriminating evaluations. There is no dualism here: The nervous system does all the processing, but its organization is a result of natural selection favoring the functionality arising from that particular organization. By favoring conscious subjective experiences that clearly discriminate between important environmental variables, natural selection, over generations, has continued to improve the neural machinery capable of generating such experiences.

What is true for the subjective quality of an experience, like sweetness or sourness, is also true for its subjective intensity: a bright light, a strong taste, a loud noise. In the external world, the actual physical attribute of a signal that changes with perceived intensity is quite different across these different sensory domains. In vision, for example, the intensity of a light is correlated with the number of photons striking the receptors of the retina, whereas the intensity of a taste varies with the concentration of molecules in a solvent, and the intensity of a sound depends on changes in air pressure over time. These various physical changes would have nothing in common except for the fact that our evolved sensory transducers convert such signals into an increased number of nerve impulses per second within the visual, gustatory, or auditory pathways, respectively, and this increased neural input evokes a subjective change in intensity. Our conscious experience, not any specific change in the physical world, defines the change in intensity of a signal. Although an increase in the energy of an environmental signal is often correlated with an increase in subjective intensity, this is not always the case. For example, a decrease in the kinetic energy of molecules striking the skin evokes an increase in subjective coldness, while a "blue" photon, which has more energy than a "red" photon, evokes a change in color rather than a change in intensity. Both the qualitative and quantitative attributes of the physical world appear to be defined, not by the events in the world that activate them, but by the evolved emergent properties of our nervous system.

The naive realism we started with—that apples appear red because they really are red—has now undergone several major modifications. According to DCS, the redness was "out there," and only symbolic representations existed, or were important, in the nervous system. The WCS view brought the redness back into the head as an emergent property. Finally, evolutionary functionalism eliminated the redness "out there" altogether, leaving it exclusively as an evolved emergent property of the nervous system. Over generations, by favoring emergent properties that enhance gene survival, natural selection has forged the neural machinery capable of generating such experiences. From this perspective, nerve cells are certainly the active agents in the nervous system, but they are organized the way they are because natural selection has favored the functional emergent properties that arise from that arrangement. It is always function—gene survival—that dictates structural design. If functional conscious experiences shaped and refined the organization of the nervous system, then the zombie we discussed earlier, with the same neural design as a conscious brain, would inevitably be conscious. At last the hard problem of consciousness has been defused, leaving us with a view of conscious experience as an active filter, or discriminant amplifier, with enormous functional benefits. This is the essence of evolutionary functionalism.

EVALUATING EVOLUTIONARY FUNCTIONALISM

Evolutionary functionalism faces the same problems as those that face astronomers: we simply have no way to replay and hence know the actual sequence of events that were responsible for creating the current state of the universe, and we have no method for reconstructing the evolutionary history of the human mind. Despite this limitation, however, two different approaches can provide strong support for the evolutionary viewpoint: computer simulation, and convincing evidence for adaptive historical design.

The first approach involves simulating the evolution of emergent properties in a computer. Computer models provide a rigorous test of any proposed origin or function of emergent attributes. If we believe, for example, that feelings evolved because they played a specific role in controlling some aspect of behavior, then it should be possible to evaluate this proposition using a computer simulation. Genetic algorithms provide a simple and effective tool for simulating the evolution of feelings in a computer; they also provide a novel method for evaluating evolutionary functionalism. The theoretical basis of such simulations, and their results, are described in Chapters 2, 3, and 4.

The second approach is to examine current emergent properties, such as feelings, to uncover their historical design. Theoretically, emergent properties evolved because they provided solutions to problems that were consistently present and posed habitual threats, or offered benefits, to biological survival in specific ancestral environments. That is, selection pressures acting over prolonged historical periods—the environments of evolutionary adaptedness (EEAs)—were responsible for the different aspects of our satisficing design. The EEA responsible for the evolution of an emergent property like "redness," for example, was certainly much earlier and quite different from the EEA that led to the evolution of a feeling like "pride." Furthermore, there is no a priori reason why an adaptation acquired in one EEA should necessarily be adaptive in another.

An evolutionary theory of feelings, therefore, should explain how each feeling evolved in response to specific problems that organisms encountered during its EEA. It should also explain why we have so many different kinds of feelings, like love, fear, and anger. This type of explanation is most convincing when it accounts for unusual characteristics that are not easily explained from alternative perspectives. As Stephen Jay Gould has persuasively argued in *The Panda's Thumb,*[9] the best evidence for historical design is the existence of characteristics that are no longer useful but that had a clear ancestral function. For example, the residual legs on a snake are an unmistakable indication of the snake's evolutionary history. In a similar manner, the existence of a feeling that had clear historical value, but is now nonfunctional or even maladaptive, would offer strong support for an evolutionary perspective. Evidence supporting the adaptive historical design of human feelings is discussed in Chapters 5, 6, 7, and 8.

THE ADAPTIVE ILLUSION

Adopting an evolutionary viewpoint on the attributes of mind solves three major problems that confront any theory that attempts to explain the nature and origin of conscious experiences. First, it resolves the "double redness" paradox by eliminating the redness in the external world, holding that all conscious experiences exist exclusively as emergent properties of neural organization. Second, it provides a functional role for these conscious experiences. They are the emergent properties of the nervous system that have been selected because they amplify and discriminate between attributes of the world that are biologically important, and as I will discuss later, these discriminations play a central role in the adaptive mechanisms of learning and reasoning. Third, it explains why sensory feelings, like sweetness or saltiness, are evoked by environmental events that are clearly important for gene survival. In the absence of an evolutionary perspective, the existence of such congruous relationships would require the assumption of a preestablished harmony between conscious experiences and biological survival.

Such preordained explanations fall outside the realm of natural science.

When conscious experiences are viewed as properties of biological tissue and not rigid properties of the "outside" world, then they can be continuously shaped and refined by natural selection. But natural selection doesn't "care" whether such experiences are accurate reflections of our external reality; it "cares" only about biological usefulness—the extent to which they enhance the survival and reproduction of organisms that possess them relative to those that do not. Limited by this single functional constraint, minds have evolved a fantastic array of emergent experiences, which in turn provide the raison d'être for neural organization. Relentless selection inexorably demands that the attributes of mind evolve to enhance discriminations when they are functionally useful, and ignore them when unimportant. As a consequence, organisms have evolved subjective experiences that impose a distorted but functionally useful view of the world "out there," and at the same time have evolved the neural machinery that underlies this interpretation of reality.

The essence of our "satisficing" design is that our senses filter out most of the background energy/matter in the world; we don't smell clean air, taste pure water, or see the vast expanse of the electromagnetic spectrum. With our receptors tuned only to specific energy bandwidths, we psychophysically scale our world, making precise discriminations over the small but biologically important ranges of intensity and frequency while minimizing or ignoring differences toward the extremes of our sensory domains. What we do detect elicits vivid conscious experiences that are gross distortions of what exists "out there." Consciousness amplifies those attributes of the physical world (and the social world, as we will see) that are biologically relevant. Our amplified and distorted picture of the world may be a powerful illusion, but it is neither arbitrary nor random; the faculties of mind were adaptively designed.

Consider a world without consciousness. The darkness is a bubbling cauldron of energy and vibrating matter, locked in the incessant dance of thermal agitation. Through shared electrons or the strange attraction of unlike charges, quivering molecules, not free to roam, absorb and emit their characteristic quantal packages of energy with the surrounding fog. Free gas molecules, almost oblivious to gravity but buffeted in all directions by their neighbors, form swirling turbulent flows or march in zones of compression and expansion, according to the dictates of oscillating substrates. A massive solar flux and cosmic radiation from events long past crisscross space with their radiant energy and silently mix with the thermal glow of living creatures, whose hungry metabolic systems pour their infrared waste into the chaotic milieu. But within the warmth of their sticky protein bodies, the dim glow of consciousness is emerging to impose its own brand of organization on this turbulent mix of energy/matter. The active filter of consciousness illuminates the darkness, discards all irrelevant radiation, and in a grand transmutation converts and amplifies the relevant. Dead molecules erupt into flavors of bitterness or sweetness, electromagnetic frequencies burst with color, hapless air pressure waves become the laughter of children, and the impact of a passing molecule fills a conscious mind with the aroma of roses on a warm summer afternoon.

In a sense, like David, we are all hallucinating. But David's conscious experiences were not in harmony with the physical world around him; his hallucinations were not adaptive. His condition may have been the result of a small chemical difference in a transmitter molecule, dopamine, or its receptor sites. This small change in brain chemistry shattered his reality. We have no particular a priori reason to value one chemistry over any other, except that our normal chemistry has been selected as part of the satisficing design that has permitted survival and reproduction on our small planet. We may not possess the optimal design for visualizing multidimensional space or warps in space-time. Nevertheless, despite their limitations, our conscious sensations and feelings are not irrelevant epiphenomena; they are remarkable emergent properties that owe their existence to that master tinkerer, natural selection.

TWO

∾

The Mother of All Codes

ON JULY 22, 1976, MY HOUSE TURNED BLACK. THIS amazing spectacle was the result of an ancient annual ritual that has probably gone on for tens of millions of years. Triggered by a rainstorm in late July, the ant colonies in the surrounding desert had released their winged males and females, and they met, by ageless agreement, on the highest landmark in the area. My house, which at that time stood all alone in the desert, had been selected for their annual primal tryst.[1] Within the almost solid black mat of writhing bodies, the ants were engaged in a much more ordinary ritual: The males were inseminating the females. Soon the nuptials were over and the females were gone, leaving only the bodies of their dead male lovers lying around the house like an eerie fossil graveyard. The males' sole function in life had been accomplished; they had passed on their precious genes.

Charles Darwin and Alfred Russel Wallace knew nothing about genes when they proposed a theory to explain how the structure and behavior of organisms could become modified over time and eventually lead to such unlikely scenarios as the nuptial flight of the desert ants. At that time most biologists were aware that the structure of animals changed over long periods of time, for such variation was directly observable from the fossil record. No one, however, had provided a good explanation of how these modifications occurred or how they could lead to the remarkably complex adaptive designs of living animals. One prominent theory, formalized by the Chevalier de Lamarck in the eighteenth century, was called the doctrine of acquired characteristics. Lamarck postulated that any characteristic of an animal that became modified as a result of repeated use during its lifetime would be passed on to future generations. If, for example, a giraffe stretched its neck in order to reach food on higher and higher branches, then this stretching would be transmitted to its offspring. In essence, the Lamarckian model of evolution was an instructional process whereby any physical or behavioral adaptation that was acquired from practice could be transmitted to the next generation.

The Darwin/Wallace model of evolution depended on selection rather than instruction. They proposed that giraffes were born with a variety of neck lengths but those with longer-than-average necks would have an advantage because they could reach the food on the top branches of trees, an option not available to their competitors. These long-necked animals would leave more offspring, who would inherit their long-neck design. In this manner the average neck length in the next generation of giraffes would be longer than the average in the previous generation. In evolution by natural selection, the structure of animals changes gradually over generations because the environment selects those that possess any functional characteristic that enhances their reproductive success.

Evolution by natural selection is possible because each member of the next generation of giraffes inherits a mixture of its parents' genes, and so it possesses the same, or very similar, genes to those that made its parents such successful survivors and reproducers. These genes contain the information for making proteins, and it is proteins that control the structure and regulate the chemistry within every cell. A small variation in the structure of genes can result in a change in the configuration of a protein, which in turn may alter the structure or chemistry within the cells of a giraffe's body. If any such modification results in a giraffe that can survive and reproduce better than those that lack this change, then this fortunate animal will contribute more of its genes to the next generation.

The Darwinian perspective makes no distinction between the structure of the brain and the structure of the body because all biological structures are ultimately organized by proteins produced from specific gene codes. As a consequence, the neural organization of an animal could be modified by natural selection in exactly the same manner as the structure of its body. Consider, for example, the peppered moth discussed in Chapter 1. In this case it would clearly be advantageous for black moths to have a behavioral preference for landing on dark backgrounds. The functional advantage of camouflage could lead to structural changes of the moth's nervous system in exactly the same way that obtaining food led to the evolution of giraffes with longer necks. Peppered moths do indeed exhibit such a behavioral preference. Natural selection modified both the animal's brain and its body because camouflaged moths enjoyed more reproductive success than those that lacked these structural and behavioral adaptations. Camouflage is an emergent property that arises from the conjoint actions of a specific pigmentation and a compatible behavioral preference. Working together, these two characteristics produced a functional emergent property that enhanced the peppered moth's survival and reproduction.

Just as natural selection can design stomachs and kidneys that possess emergent properties like digestion and excretion, it can also design brains that possess such emergent properties as sensations and feelings. That is, if an animal experiences a discrete feeling, such as sweetness, when it eats a valuable resource and a different feeling, bitterness, when it encounters a toxin, then it will enjoy the benefits of this ability to form such discriminating evaluations. If such emergent properties increase the animal's survival and reproduction relative to animals that don't possess them, then any genes that contribute to this design will increase in frequency in future generations. Under such circumstances a host of complex emergent properties can evolve and become highly refined over generations.

For animals adapting to their environment, evolution by natural selection is a parallel process. Some animals may survive' because of their longer necks, others because of their efficient digestive enzymes, and still others because of feelings that permit them to discriminate between nutrients and toxins. Since the genes that contribute to any of these survival-enhancing features enjoy more representation in subsequent generations, their frequencies will continue to increase, and the resulting adaptive characteristics will become more and more widespread. Evolution, then, is simply defined as a change in gene frequencies over generations, and natural selection appears to be the major factor that is responsible for directing this process. More than one hundred years of study have shown that the selection procedure works, while there is no convincing evidence for Lamarck's theory of the transmission of acquired characteristics.

The insights of Darwin and Wallace were remarkable, given the fact that they had no knowledge of genes or how they could influence the structure of organisms. Today, however, we do possess this knowledge and can use it to simulate evolution in a computer. As I will demonstrate in Chapter 4, we can even use such simulations to examine how our feelings could have evolved. First, though, we need to understand what genes are, how they regulate development, and how they are modified and shuffled during sexual reproduction. Let us begin by examining the structure and function of proteins.

MESSAGES FROM INSIDE THE SOMA

Proteins are the crucial elements that control the form and function of each cell of a living organism. These large macromolecules can be structural elements that endow cells with important physical properties, or they can be enzymes that regulate the chemical reactions within cells. The differences among kidney cells, liver cells, and nerve cells are a consequence of the different sets of proteins that are being produced within these various types of cells and the cellular chemistry that depends on the actions of these proteins. Yet despite their many complex roles, all proteins have a very simple underlying design. Every protein is a unique sequence of amino acids joined together in a long chain. The specific order of these amino acids is called the *primary structure* of the protein. The complexity of proteins arises from the fact that their length typically runs for more than two hundred amino acids, and each amino acid may be one out of twenty possible varieties. As a result, the total number of possible proteins is greater than twenty raised to the power of two hundred (20^{200}). It is difficult to grasp the enormity of this number. To say that it is astronomical is a gross understatement. If we made just one copy of every possible protein we would have billions of billions of times more protein molecules than there are atoms in the entire known universe! For cells to function correctly, however, they must be able to make a very specific set of proteins out of this gigantic number of possibilities. The ability to precisely specify the unique primary structure of a specific protein is the remarkable property of a gene.

A gene is a section of DNA that specifies the sequence of amino acids in a particular protein. Like proteins, the underlying structure of DNA is quite simple. It can be visualized as a very long ladder that has been twisted so that its sides form two parallel corkscrews held together by the rungs or steps of the ladder. The two sides of the ladder are made of an alternating sequence of a simple sugar, deoxyribose, and a phosphate group (Figure 2.1). This alternating sugar-phosphate-sugar-phosphate sequence is identical in every piece of DNA. The rungs of the ladder, however, vary along the length of the DNA molecule, and it is they that are responsible for designating the sequence of amino acids in a particular protein.

Each rung is made from two bases, and each base may be one of four possibilities: adenine (A), guanine (G), cytocine (C), or thymine (T). A complete rung, therefore, consists of a single base that is firmly attached to a deoxyribose group on one side of the ladder and a second base that is firmly bonded to another deoxyribose group on the other side. The two halves of the DNA molecule are then connected by the weak hydrogen bonds that hold the bases together. For example, an A on one strand can attach to a T on the other strand using two hydrogen bonds, and a C on one strand can form three hydrogen bonds with a G on the other strand. Other combinations of bases do not fit together, so if the base sequence on one strand of the molecule is known, then the sequence of bases on the other strand is also known; it must be the complementary sequence of bases. This important feature of the DNA molecule allows it to replicate itself every time a cell divides into two cells.

During replication, the DNA molecule unzips by breaking the weak hydrogen bonds between the complementary base pairs. Each exposed half of the molecule then serves as a template for constructing a duplicate copy of the half that was lost. This remarkable property can be demonstrated in a laboratory using an elegant process known as the polymerase chain reaction (PCR). During PCR, a solution containing a double-stranded piece of DNA is repeatedly heated and cooled in order to break and reconstruct the hydrogen bonds between the strands. In the presence of small complementary DNA sequences that can attach to the ends of the

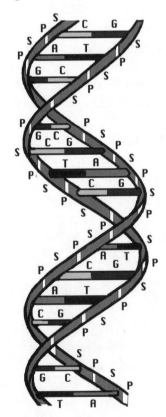

Figure 2.1: The twisted ladder of DNA showing the sugar-phosphate sides (S and P) and the base pairs that make up the rungs (A, T, C, and G).

separated strands and prevent them from recombining, together with the appropriate enzymes and building blocks, each heating-cooling cycle will double the amount of DNA that was present in the prior cycle. In this manner, many millions of copies of a piece of DNA can be generated within a very short period of time.

In laboratories throughout the world, multiple copies of pieces of DNA taken from the tissue of living or dead organisms are now produced using PCR. These copies can then be used to decipher and compare the sequence of bases found in the DNA of different plants, humans, or other animals. As a result, the entire base sequences of some plants and animals have been entirely deciphered. A complete sequence map of the 3.3 billion bases that make up human DNA, it is estimated, will be known sometime early in the twenty-first century. The value of this knowledge is enormous since it is the base sequences of genes that ultimately determine the primary structure of the proteins manufactured from those genes.

The synthesis of a protein from its gene code occurs in two distinct steps: *transcription,* followed by *translation.* The gist of transcription is that a gene segment of DNA unwinds, and a strand of messenger RNA is then built on the exposed DNA template. RNA is similar to DNA, but it is a single-stranded molecule with the sugar ribose replacing deoxyribose, and the base uracil (U) replacing thymine (T). During transcription the base sequence on one side of the DNA, the so-called sense side of the molecule, is copied by using the exposed bases as a template for building a single-stranded RNA molecule that has the complementary sequence of bases. This messenger RNA then leaves the nucleus of the cell, where, with a great deal of assistance from enzymes, ribosomal RNA, and various forms of transfer RNA, its base sequence is translated into the amino acid sequence of a specific protein.

When a protein molecule is produced, it does not remain as a long loose thread. Rather, as a result of the interactions between the amino acids of the primary structure, it rapidly adopts a secondary structure. This secondary structure exposes portions of the molecule to its surroundings, and its interaction with chemicals in this local environment determines the final, or tertiary, configuration of the molecule. The important properties of proteins are a direct consequence of their ability to rapidly change their tertiary configuration in response to different environmental signals. For example, one gene might prescribe the primary structure of the protein actin, and another gene might code for the protein myocin. These two large proteins are produced inside skeletal muscle cells, and the contraction of muscles depends on making and breaking the bonds between them. Muscles contract in response to chemical events initiated by messages from nerve cells. These chemical signals cause the cross bridges from myosin onto actin to change their shape, and these "rowing movements" pull the proteins past each other, resulting in muscular contraction.

Humans apparently can make about a hundred thousand different proteins. The codes for the primary structure of all of them are stored as genes on the forty-six segments of DNA, the chromosomes, that are found within the nucleus of the cells that make up a human body. Because DNA is duplicated every time a cell divides, the DNA within each cell of an individual's body is an exact copy of the original DNA inherited from the individual's parents. Exceptions occur during the production of sperm or ova, when a particular kind of cell division reduces the amount of DNA by fifty percent, in short-lived red blood cells that don't contain DNA, or as a consequence of somatic mutations that can result in small changes to the structure of DNA, an effect that is of particular importance within the cells of the immune system.

Apart from these relatively rare occurrences, the protein codes written on the DNA molecules within any individual are identical in every cell of his or her body. That is, the forty-six chromosomes within every cell of a person are an exact copy of the genes donated by that individual's parents. There is thus little doubt that DNA is passed from generation to generation and that this genetic material, found within the nucleus of every cell, contains the important information required for manufacturing the proteins that regulate the structure and function of every cell of a living organism. The consequences of this arrangement are quite startling. If the nucleus of a fertilized ovum is replaced with a nucleus taken from an adult cell, then it is possible to generate an exact genetic duplicate, or clone, of the adult donor.

In 1997 Ian Wilmut and his colleagues described how they

were able to produce a mammalian clone, Dolly, using a nucleus taken from an adult sheep. This experiment raised the possibility of creating human clones and generated a political, cultural, and scientific uproar that will undoubtedly resonate within society for many years to come. Cloning, however, is not a new or unnatural process. It is, after all, merely an artificial form of asexual reproduction, which is the most common method of reproduction on the planet. Most microorganisms, like bacteria and yeasts, reproduce by making clones of themselves. For example, bacteria replicate their DNA and then divide, so that every generation is genetically identical to the previous generation. They are, in essence, clones. But if this were the whole story, bacteria would never evolve, since their genes would remain constant for generation after generation. This is not the case, however, because a number of processes—such as mutation, conjugation, transformation, and transduction—invariably generate variations of bacterial genes. Because of their rapid reproduction rate and these rich sources of genetic diversity, bacteria and other microorganisms can evolve very quickly. This presents an enormous problem for large multicellular organisms, like humans, who have to cope with the continually changing biological threat.

For any multicellular organism, the greatest danger to survival has been, and continues to be, invasion by parasitic microorganisms like bacteria, protozoa, viruses, or fungi. Indeed, it has been estimated that about half of all humans who have ever lived on this planet were killed by the plasmodium protozoon, which causes malaria. This single disease continues to kill about two million people every year, one individual every fifteen seconds. In fact, several times as many humans die each year from malaria as died in the recent genocide in Rwanda. Bacteria can also be lethal. In the sixth century a major outbreak of the pneumonic plague, caused by the bacterium *Yersinia pestis,* killed more than 100 million people as it swept across the Middle East, Europe, and Asia. A second pandemic in the fourteenth century killed almost half the population of Europe, about 75 million people. Viral infections are also a major cause of human death. In 1918, following World War I, a particularly virulent strain of influenza claimed more victims than the four-year war. The harmfulness of these parasitic invaders rests on their ability to rapidly evolve proteins that can bind to the cell membranes of their victims, enter their cells, and use the captured resources to make further copies of themselves. Successful invaders then multiply rapidly and pass on their DNA to the next generation. Inevitably some offspring will be better binders than their parents, and an epidemic is soon under way. This enormous problem confronts all multicellular organisms.

To cope with invading parasites, large animals and plants must constantly vary the structural proteins of their cell walls in order to retard the evolution of effective binders. To remain constant over generations is a suicidal strategy. The Irish potato blight is a lasting monument to the fallacy of reducing genetic variation. In the Peruvian Andes there is great genetic diversity in the wild potato population, but early Spanish explorers of South America brought back only a small number of potatoes to Europe. The Irish potato plants were all descended from these few strains. They lacked genetic diversity, and when the fungus *Phytophthora infestans* attacked the Irish crop in the 1840s, the blight was devastating: More than 1.5 million people died of starvation. For a long-living animal or plant, the solution to evading rapidly evolving parasites is to ensure that its offspring are somewhat genetically different from itself. This appears to be the major reason for the evolution of sexual reproduction, and it is the strongest argument against the widespread use of human cloning.

Most long-living multicellular organisms practice some form of sexual reproduction. In humans, for example, the first cell of an offspring contains twenty-three chromosomes that were donated by an ovum and twenty-three that were provided by a sperm. This means that at conception the fetus has two copies of every chromosome and, of course, two copies of every gene that is found on these chromosomes. These homologous or paired chromosomes contain one copy of each gene that came from the individual's father and one copy that came from the individual's mother. Each individual is therefore a new mixture of his or her paternal and maternal genes. This, however, is only part of the story. When the father and mother were forming their germ cells, sperm or ova, they mixed together the genes that they had received from their own parents. That is, during the synthesis of germ cells, each chromosome within the father (or mother) exchanged pieces of its DNA with its homologous partner, and only one of these new commingled versions was placed in each sperm (or ovum). In this manner the number of chromosomes in the germ cells was reduced from forty-six to twenty-three. At the same time the "crossover" process resulted in chromosomes that contained completely new combinations of genes. If crossover did not occur, then every new individual would be a mixture of his or her parental chromosomes. Because of crossover, however, whole chromosomes do not maintain their integrity from generation to generation, so each offspring is a mixture of its parents' genes rather than their chromosomes.

Compared with chromosomes, genes are very stable and may remain intact over many, many generations. Despite a host of mechanisms for checking and repairing errors, however, genes sometimes get transposed, inverted, deleted, duplicated, or changed. For example, ultraviolet (UV) radiation from the sun has been, and continues to be, a major source of damage to the DNA molecule. A photon of UV radiation is particularly destructive if it strikes a T base that is adjacent to another T on the DNA molecule. Under these conditions the two thymines may react to produce a photodimer. If such a mutation occurs on a segment of DNA that produces an important protein, it could have fatal consequences. As you might expect, however, damage from UV radiation has always been a constant threat to biological survival, and sophisticated repair mechanisms have therefore evolved to cope with such damage. In this case an enzyme (UV-endonuclease) can recognize the dimer, excise the error, and replace the missing bases using the information available on the intact complementary strand. Sometimes, however, a mutation can damage the repair mechanism itself, producing a disorder known as *xeroderma pigmentosa*. In this instance, exposure to sunlight will produce ubiquitous sores and eventually life-threatening malignant tumors. Most mutations are destructive, but on rare occasions mutant genes in a sperm or ovum can have favorable consequences for the survival or reproduction of offspring. Such mutations can influence the course of evolution.

In sexually reproducing species, crossing-over and mutations appear to be the major factors that generate genetic diversity.

During reproduction these variations in gene structure are passed from generation to generation. Darwinian evolution is a consequence of these random[2] variations in gene structure as well as the nonrandom selection of organisms on the basis of their functional characteristics. These are the basic principles of evolution (and will be simulated by a computer model described in the following chapter). Genes, however, do not act alone in producing adaptive designs. As an individual develops from a single fertilized cell into a multicellular organism, cell differentiation occurs because genes within particular cells are turned on or off according to the dictates of environmental signals. Such signals, which originate in either local or remote locations, may take the form of proteins generated from other gene codes, chemical messages from local cells, blood-borne hormones from distant locations, products resulting from the action of other enzymes, or even chemicals from the maternal environment—that is, chemicals originating outside the body that is under development.[3] These chemicals can influence gene expression by changing the tertiary structure of binding proteins that are attached to regulatory elements of the DNA molecule. Regulatory elements are segments of DNA, usually within 500 to 5,000 bases from the boundary of a specific gene, that govern when, where, and how a gene is activated. Changes to the structure of binding proteins on regulatory elements determine when a gene is turned on or turned off during development, the tissue type in which the gene is active, and the amount of protein produced by that gene. The constant interactions between genes and the environmental signals to which they are exposed regulate the structure of every aspect of a developing organism, including its nervous system.

MESSAGES FROM OUTSIDE THE SOMA

The biological design of any individual begins at conception and continues until death. Major changes to the body and brain occur during the early embryological stages and again at puberty, but the basic developmental processes of replication, differentiation, modification, repair, and replacement never cease. Although entire cells of the brain and spinal cord are not replaced as whole entities, they are continually repaired and modified by gene-environment inter-actions. In a fully developed individual, external environmental events originating outside the organism are converted into com-plex patterns of chemical messages released by nerve cells onto their neighbors. These neurotransmitter signals (often through the action of second messengers) continue to alter the expression of genes and consequently modify the structure of the brain through-out an organism's lifetime. Neuronal gene expression is constantly regulated by its chemical environment, just as all cell differentia-tion has been regulated since the moment of conception. But for a nerve cell tucked away in some corner of an individual's brain, distant events from the external world, translated into a surge of chemical messages released by its local neighbors, now modify its gene expression. The sphere of influence acting on each gene has become extended over the course of the individual's development. Each individual's personal experiences awaken and regulate the broad developmental plan that is stored in archives of DNA and handed down from generation to generation. In this manner both the "pre-wiring" of the brain during development and the "re-wiring" of the brain as a result of an individual's environmental experiences depend on the structural modification of neural net-works achieved by genes responding to specific chemical signals in their environment. That is, both *phylogenetic learning,* achieved through natural selection over generations, and *ontogenetic learning,* arising from an individual's unique experiences, share this common mechanism of neural modification mediated by protein synthesis. (In Chapter 4 the basic evolutionary model described in Chapter 3 is developed into a more complete computer model that simu-lates both the ontogenetic and phylogenetic learning mechanisms.)

Given all that we have learned about the interaction between genes and environment over the last fifty years, it is difficult to understand why arguments still persist over whether a particular behavior is the product of one factor or the other; it is always both. Nature and nurture are the inseparable twins of development—one cannot exist without the other. No behavior is immune to genetic influences, since they are responsible for the development of the very nerves, muscles, and glands that make behavior pos-sible. The same genes, however, may be expressed differently in different environments. For this reason, it is not possible to specify how much of a behavior is due to one factor and how much to the other, since all behaviors are totally dependent on both. A series of simple thought experiments will clarify this issue and, at the same time, explain the interaction between nature and nurture in determining human behavior.

Consider two genetically identical seeds that are grown under the same environmental conditions of light, temperature, nutrients, and so on. During development, however, one plant receives twice as much water as the other, and as a consequence it grows six inches taller. The height of the first plant is clearly a function of both its genes and its environmental experience; the same is true for the height of the second. But in this case the difference in height is exclusively a consequence of the plants' different environmental experiences. Since the plants are genetically identical, and each could have benefited equally from the additional water, the difference between them can be attributed only to the differences in their environments: the fact that one received twice as much water. Now consider two genetically different seeds grown in identical environments. They receive exactly the same amount of water. In this case one plant may still grow six inches taller than the other. Again, the height of each plant is a consequence of both its genotype and its environmental experience, but in this case the variation in height is exclusively a consequence of a difference in genotype. These two examples demonstrate that differences in a phenotypic characteristic, such as height, can be due to differences either in genes or in environmental influences. But a third factor, the interaction, must also be considered.

Now consider two pairs of seeds; each pair has an identical genotype but a different genotype from the other pair. During development one member of each pair receives more water than its identical twin. In this case all the plants may end up with quite different heights, but as always, the height of each plant is still a product of both its genes and its environmental experience. Note, however, that although the additional water given to a plant of one genotype may cause it to grow six inches taller than its twin, the same additional amount of water may result in a mere two-inch height difference between the plants of the second genotype. There is, in this case, an interaction between genes and environment, since the same environmental influence has a different impact on plants with different genotypes. Using groups of plants raised under these different conditions, it is possible to experimentally determine how much of the total variance in height between plants is due to differences in genes, differences in environmental experience, and differences resulting from gene-environment interactions. It is not the height itself but the variation in height between plants that can be attributed to these different factors.

What is true for height is true for any phenotypic feature, including a behavioral attribute. Although we can never specify how much of a particular behavioral characteristic, like aggressiveness, is due to either genes or environment, it is possible to empirically determine what proportion of the difference in aggression between individuals is due to differences in their genotypes, their environmental experiences, or the interaction between their genes and their experience. That is, it is possible to divide the variation in any structure or behavior (phenotypic variance) into variance of genetic origin (G), variance of environmental origin (E), and variance due to the interaction between genes and experience (G × E).

$$\text{Phenotypic Variance} = G + E + (G \times E)$$

Without experimentation, it is simply not possible to know what proportion of the variance in any behavior can be attributed to these different sources. Behavioral geneticists often use twin studies to compute the proportion of phenotypic variance that can be attributed to genetic influences (the heritability coefficient), but practically speaking, psychologists are often more interested in the proportion of variance that can be attributed to environmental factors—the observed "plasticity" of a behavior. Should we expect all behaviors to be equally modifiable by environmental experience? May some underlying considerations explain why a high degree of environmental plasticity is found for some behaviors, whereas, for others, low environmental variance is common? We can gain some insight into these questions by considering the different types of problems that organisms encounter when facing the problems of surviving and reproducing on this planet.

Animals inhabit complex ever-changing environments. Some environmental features—for example, the nature and location of food—may undergo very rapid change; other features, such as the nature of local predators, may change more slowly; and other features—like the appearance of other members of the same species and the length of the day-night cycle—remain constant over many generations. That is, environmental factors change at different rates. When the time between generations is short compared with the rate of change of the environmental features with which the animal is interacting, natural selection alone provides an adequate mechanism for achieving structural and behavioral adaptations. Most organisms, for example, have evolved endogenous circadian rhythms that exploit the relatively stable length of the day-night cycle and postural mechanisms that exploit the stable direction and strength of gravity. The sexual behavior patterns of many organisms are responsive to the physical appearance of the opposite sex—an environmental constancy of biological origin. Since these environmental factors remain constant from generation to generation, it is not necessary, or adaptive, for organisms to possess a high degree of behavioral plasticity to effectively interact with them (Figure 2.2). Indeed, a high degree of environmental variance could be dangerous; one might fall in love with an orangutan!

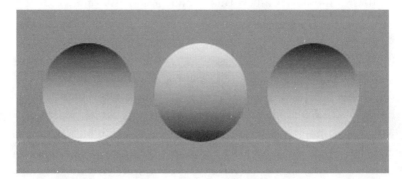

Figure 2.2: The implicit assumption of an overhead sun causes us to perceive the middle circle as a convex surface and the outer circles as concave surfaces. Try inverting the picture.

But organisms with relatively long lifetimes require new behavioral mechanisms in order to adaptively interact with more rapidly changing aspects of their environment. Learning and reasoning are two important adaptive processes that allow humans and other animals to modify their behavioral interactions as a function of their unique experiences. Learning allows animals to acquire and store new behaviors for interacting with aspects of the environment that vary in a consistent manner within their lifetime. Reasoning empowers them with the ability to predict future changes based on this stored information. Such complex adaptive processes, however, can evolve only if they contribute to the biological survival of the organisms that possess them. To be adaptive, organisms must learn and reason about those circumstances that benefit or threaten their gene survival, and it is feelings that signal which circumstances may do one or the other. How feelings evolved, and how they now serve as rewards and deterrents that regulate learning, is discussed in Chapter 4. But first we must explore how to simulate evolution in a computer.

THREE

∾

Searching FaceSpace

ON THE MORNING OF FEBRUARY 11, 1990, THE HEAD-line of my hometown paper, the *Las Cruces Sun News,* blazed, "4 Killed in Massacre." The previous Saturday a father, his two daughters, and another girl were killed during a bowling alley robbery, and local police were calling it the worst massacre in more than twenty years. The details of the case were gruesome. The two killers had herded seven people, including a two-year-old child, into an office. In order to conceal their identity and destroy all evidence of their paltry $5,000 theft, they shot all seven people in the back of the head and then set fire to the premises. Fortunately, there were three survivors. The local police, however, were faced with the formidable task of trying to determine the identity of the killers based on the memory of traumatized victims and several casual eyewitnesses who happened to be near the scene of the crime. Apparently they had no other evidence.

Over the next few days the facts of the case became clearer, yet the problem confronting the police grew in magnitude. Most of the surviving victims had sustained severe head injuries, which virtually always produce some degree of retrograde amnesia. While a memory is in the process of being turned into a permanent structural modification of the brain, as a long-term memory, it exists for several hours in the vulnerable state known as short-term memory. Such memories are easily disrupted by a number of physical and chemical processes, including a trauma to the head. Typically such memory loss, which is very common following motorcycle accidents, manifests itself as a failure to remember events just prior to the trauma, even while long-term memory remains intact. In the bowling alley case all of the survivors had been shot in the head, so only incidental witnesses could provide reliable descriptions of the murderers. If even one of the assailants had had a prior criminal record, one of these witnesses might have been able to make an identification from mug shots. It soon became clear, however, that this was not the case, and the police were forced to build facial composites based on the eyewitnesses' descriptions. Therein lay the heart of the problem.

Humans have very poor facial recall ability, yet all current methods of generating facial composites—from the use of sketch artists to sophisticated computer programs—rely on the ability of witnesses to accurately recall the face or facial features of a culprit. To comprehend this difficulty, imagine that you have to describe the face of a close relative—your mother. How wide is her nose? How thick are her lips? What is the shape and slope of her eyes? The task seems close to impossible, and looking at isolated features doesn't make the problem any easier. Imagine trying to select your mother's nose from a whole array of noses. "Well, which one is it?" asks the detective. "I think it's the average one" is the common reply!

Now consider a remarkable fact. Suppose you are casually introduced to someone at a party. The following day, you meet that person on the street. You glance at each other, say hello, and at that moment each of you knows that you have met before, although you may not remember where or when. Unlike facial *recall,* we are all experts in facial *recognition.* After unsuccessfully attempting to describe your mother's face, you may be wondering if you really know what she looks like. Don't worry! We effortlessly recognize our family members whenever we see them, and we can easily identify them from pictures. This ability is not confined to close relatives. We identify a large number of friends by their distinctive facial characteristics, and we can often recognize a fellow student or even a casual acquaintance whom we haven't seen for many years.

During our daily conversations human faces are our major focus of attention. These small areas of skin are the most scrutinized territories on the entire planet. It is not surprising, therefore, that faces contain the most important cues for identifying individuals and for discriminating between strangers, kith, and kin. Without the ability to identify individuals, our social lives would be chaotic and even dangerous. The human brain appears to have a distinct bias for faces, since as early as at three weeks of age, babies will look at a face in preference to any other visual image. By eight weeks they can discriminate between their mother's face and the face of a stranger, and they soon show a noticeable preference for familiar faces. All of these observations suggest that specific neural mechanisms are devoted to facial processing, as does the observation that as many as ten percent of the nerve cells in the inferior temporal cortex of a primate brain respond to pictures of faces. In humans bilateral lesions of this area can result in an unusual clinical syndrome known as *prosopagnosia,* literally "not knowing people." Individuals suffering from prosopagnosia can not recognize close family members and may even fail to recognize their own face in a mirror! Even stranger, the disorder is not accompanied by any loss in visual acuity or memory failure. Prosopagnosia patients can immediately recognize individuals as soon as they speak, they can describe faces in detail, and they can even match front-facing faces with profiles of the same individual. What has been lost, however, is visual recognition ability, the ability to associate the face with the identity of the individual.[1]

We may not be able to recall a face, or even a single feature of that face, but when presented with the correct set of cues, we may experience instant recognition. Somewhere in the depths of our brain a detailed memory is ready to be awakened by the appropriate visual input. Although an eyewitness may not be able to describe a criminal, he or she might very well recognize the culprit if presented with the right visual cues. At the time of the bowling alley murder, however, no method of presenting such cues existed. As the case progressed and the police composites appeared in the local paper, a sense of hopelessness pervaded the town. The composites had been built from recall, not from recognition, and it was painfully clear that they were inadequate (see Figure 3.1).

As I sat examining the initial composites over breakfast one morning, my mind was transported back to a conference I had attended three years earlier at the Santa Fe Institute. John Holland, a professor of psychology and computer science at the University of Michigan, was explaining how the principles of evolution could be simulated in a computer using a genetic algorithm. During his presentation it had occurred to me that the same procedure could allow a witness to "evolve" a culprit's face in a computer, using recognition ability alone. As he talked, I imagined random computer genes generating many different faces on a computer screen, and I visualized a witness pointing at the ones that resembled the culprit. These faces, of course, were only slightly more like the culprit than the other faces, but like giraffes with slightly longer necks, they could breed to produce offspring that were a mixture of their genes. As I listened to the lecture, I put the pieces together. Giraffes with longer necks survived and had more offspring than those with shorter necks; in the same way, faces that were closest to the culprit would have more "offspring" than those that bore less resem-

Figure 3.1: First police composites of the Bowling Alley murderers

blance to the culprit. The witness could use a scale to rate the faces, the numbers 0 to 9 on the keyboard. The fittest faces would be given a higher rating, and their "children" would "inherit their genes." The next generation would be better images because they were the "offspring" of the best faces in the prior generation. I found myself getting more and more excited. It wouldn't even matter if the witness knew why a particular face resembled the culprit; all they had to do was give it a higher rating. As Holland explained how long strings of 1s and 0s could simulate genes that could cross over and mutate between generations, I designed a way for these "genes" to make faces. By the time the lecture was over, I was sure that I had invented a new way of "evolving" a culprit's face, and I was convinced that it would work.

Three years had passed since the Santa Fe conference. Now at last the criminal justice department had provided some funds, but it was too late for the Bowling Alley case. The murder still disturbed me. Since the initial composites were so poor, I became more determined than ever to turn my idea into a reality. I developed and tested FacePrints over the next few years, and it later proved to be a pivotal tool for understanding the biological significance of our human notion of beauty.

TRACKING A CRIMINAL THROUGH FACESPACE

The simplest way to build a facial composite using recognition ability (rather than recall) is to formulate the problem as a search for a target face—the culprit's face—within a very large multidimensional "face-space." To begin, consider a three-dimensional space like a box. Now imagine a large number of hairstyles arranged along the front edge of this box beginning with bald heads on the left and progressing to very long hairstyles on the extreme right. This is the X-axis of the face-space. The Y-axis, from bottom to top, begins with very pointy chins at the bottom and ends with broad double chins at the top. Finally, from the front of the box toward the back there is a Z-axis; this is the ear dimension that begins with small flat ears at the front and progresses in an orderly fashion to large protruding ears toward the back. Using this three-dimensional face-space, it is possible to specify the outline of any face using only three coordinates. For example, the coordinates 256, 12, 97 would correspond to hairstyle 256, chin type 12, and ear type 97. Indeed, every point in the three-dimensional space corresponds to a unique facial outline that can be described by only three numbers.

Now mentally increase the number of dimensions of this face-space to include noses, mouths, eyes, and eyebrows; a total of seven dimensions so far. Further expand the face-space to include the distances between features as additional dimensions. These include the distance from the eyes to the hairline, nose, mouth, chin, ears, and eyebrows, as well as the distance between the eyes. In practice, these proportions are best represented as displacements from the average. For example, a value of minus 4 on the eye-eye dimensional axis would be interpreted as four pixels less than the average eye-eye distance. Although we can't visualize such a multi-dimensional face-space, any single point in it is now a complete face that can be defined mathematically as a sequence of only four-teen decimal numbers corresponding to the values on each of the fourteen spatial coordinates. If each decimal number is now converted into binary code (like 1100100010 . . .), then a complete face, including all of its proportions, can be written as a long string of 1s and 0s. The length of this string can be expanded to cover as many values of features and proportions as desired. As our program developed, we found that a 60-bit string was sufficient to cover an enormous range of male Caucasian faces.

A good analogy is to view every 60-bit binary string as a genotype made of fourteen "computer genes" that specify the features and proportions of a face. A computer program can easily decipher any genotype by decoding these genes, selecting the appropriate features from a database, and correctly positioning them on the computer screen. In my Macintosh computer it takes less than one second for this "developmental" procedure to convert any 60-bit genotype into a photographic-quality image of a face, a phenotype. Using the 60-bit strings, the program can generate more than a billion billion different faces (2^{60}), a number that is millions of times larger than the current world population. With this model in mind, the task of a witness—to generate a facial composite—now becomes to search this enormous face-space in order to find a single point: the culprit, or a close resemblance to the culprit. This is not a simple task. If a witness made a linear search through all the faces, viewing one face per second, night and day, it would take more than 36 billion years—several times the age of the universe! The process in question, however, uses an interactive genetic algorithm to make a highly efficient nonlinear search of this virtual face-space. This software program, called FacePrints, can find a close resemblance to the culprit in less than one hour.

The FacePrints computer program begins by creating thirty random genotypes, each 60 bits long. This "first generation" of facial genotypes is, in effect, thirty random points scattered throughout the multidimensional face-space. One at a time each genotype is developed into a face and displayed on the computer screen. The witness is simply instructed to examine each face, decide if it has any resemblance to the culprit, and register the decision using the numbers 0 through 9 on a computer keyboard. A low rating, 0 or 1, signifies that the face has little or no resemblance to the culprit, whereas a 9 means that the face is perfect. Since the faces in the first generation are random, they will most likely receive very low "fitness" ratings, almost always less than 3.

Now comes the exciting part. After these "fitness" ratings have been collected for the first generation of thirty faces, a selection procedure assigns the genotypes for breeding the next generation. This works like a raffle, in which each genotype has a number of tickets equal to its fitness rating. The raffle tickets are randomly drawn, two at a time, and the two selected genotypes are then permitted to breed. That is, the computer model selects and breeds faces in proportion to their fitness rating. Breeding employs two procedures: crossover and mutation. When any two genotypes breed, they exchange random portions of their genotypes (Figure 3.2), so that their offspring are random mixtures of the parental genes. During breeding there is also a small probability of a mutation. A mutation is simply a 1 on a genotype changing to a 0, or a 0 changing to a 1. Following selection and breeding (with crossover and mutation), the two offspring genotypes are developed into faces and rated by the witness, one at a time. If either offspring receives a rating that is higher than the least fit member in the current population, then the latter "dies" and its genotype is replaced

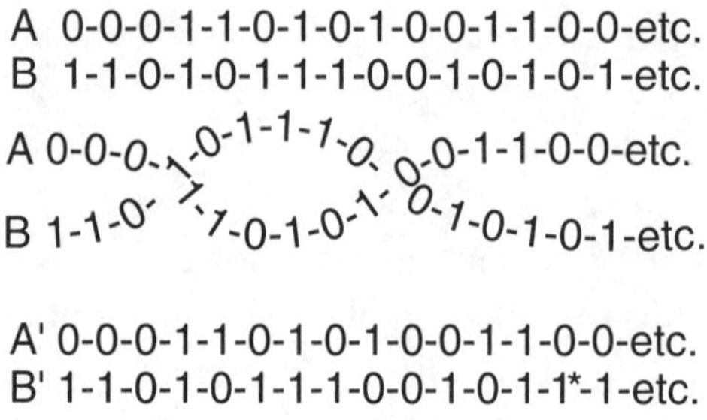

A 0-0-0-1-1-0-1-0-1-0-0-1-1-0-0-etc.
B 1-1-0-1-0-1-1-1-0-0-1-0-1-0-1-etc.

A 0-0-0.~0-1-1-1-0. 0-0-1-1-0-0-0-etc.

B 1-1-0´ ´1-0-1-0-1´ 0-1-0-1-0-1-etc.

A' 0-0-0-1-1-0-1-0-1-0-0-1-1-0-0-etc.
B' 1-1-0-1-0-1-1-1-0-0-1-0-1-1*-1-etc.

*Figure 3.2: Crossing-over of two genotypes (A and B) to produce two offspring (A' and B'). The * denotes the location of a mutation.*

by the offspring's genotype. Replacement of the worst member ensures that the population size remains constant. As selection, breeding, and the evaluation of offspring continue, generation after generation, the average fitness of the population steadily increases. The process continues until the witness concludes that a satisfactory composite of the culprit has been "evolved."

Imagine yourself in the position of a rape victim, sitting in front of a computer screen, attempting to recreate the face of your assailant. At first the faces have little or no resemblance to the culprit, and you feel annoyed by the whole process. Some of these faces are "really ugly," you think to yourself, but they're not your monster. You rate them all 0. You're beginning to wonder what's going on, but reluctantly give a rating of 1 to a face "of about the same age" or a 2, because he has "approximately the same amount of hair." Still, your directions appear to be ignored, and you are forced to reject face after face with 0 after 0. By face 50, the rejections are getting tedious, and you wonder if the program is working at all. "Maybe I'm doing it wrong," you think. "Maybe I didn't understand the instructions." You wonder if you have wiped the whole event from your mind, losing the face forever. But you reluctantly persevere. Every now and then you see a face that feels right in some vague or distant way that you can't quite discern; you give it a higher rating.

You have no awareness of computer genes or generations, but in some uncanny way the faces appear to be getting a little better now. "That one deserves a 5," you think. Later you exclaim, "Now I remember—those piercing eyes!" For a while it looks hopeless again. Face 143 appears, and you murmur, "That's like him," but you quickly hope that no one heard you, since 144 looks like a mutant from a distant galaxy. At 152 you exclaim aloud, "There he is!" then just as quickly follow it up with "No, wait a minute, there's something wrong." At least it's more exciting now. The faces do look like him. Each one is a variation of the one before. Face 169— they're actually quite good now. You find that you are holding your breath. It's a little scary. It's working. He might be in the next pic- ture. Then suddenly, at 183, the cues are powerful. "We've got him!" you exclaim with a surge of fear and excitement—"But the lips were bigger." Almost as if the computer heard your trailing complaint, number 184 has bigger lips. "Bingo!" you scream. "I'm sure of it!" You reach out and press the 9 key, and immediately fall back in your chair with a feeling of satisfaction. There on the com- puter in front of you is a photograph of the monster who raped you, and just below it is a long sequence of 1s and 0s: a unique code, a "fingerprint" that the police will be able to use to search their stored records. You did well, and you know it.

As you can see, the genetic algorithm originally conceived by Holland is a highly efficient search procedure.[2] FacePrints is an example of an interactive genetic algorithm where a human—the witness, in this case—acts like a complex environment and sup- plies the measure of fitness associated with each phenotype. Interactive genetic algorithms are very useful for designing systems to meet the needs of human operators. The operator may give feed- back in the form of a behavioral performance measure, a physio- logical response, or an aesthetic preference. For example, a genetic algorithm could be used to uncover the best configuration of a computer keyboard in order to optimize typing speed, or the lay- out of a complex display panel in order to meet the demands of a user.

Designing aesthetically pleasing products for humans or other animals is another exciting application of interactive genetic algo- rithms. Richard Dawkins, for example, has suggested a method for finding the visual attributes of a flower that are most attractive to bees.[3] In this case the binary genotypes would dictate the size, color, and configuration of computer-generated virtual flowers. A first generation of random flowers would be displayed on a computer monitor that would then be placed in a garden, in close proximity to a beehive. A touch screen could count the number of bees that landed on each phenotypic flower. After a period of time, or perhaps daily, a new generation of flowers would be evolved, using the count as a measure of fitness of each flower. Over several weeks this procedure should, at least in theory, uncover the flower that bees find to be most attractive! As we will see later, a similar procedure can be used to uncover what humans find most attractive in other humans.

A number of variables, such as mutation rate and rate of crossover, influence the efficiency of a genetic algorithm. If mutation rate is too high, then successful genotypes are easily destroyed. If the rate is too low, then variability may be insufficient to find the best answer to a problem. One solution to this dilemma is to use a metalevel genetic algorithm in order to evolve the optimal values for these two parameters.[4] The probabilities of mutation and crossover, at any point on a genotype, can themselves be coded on a binary string, and their optimal values can then be evolved over generations. In FacePrints, for example, it is possible to find the best values for mutation and crossover rates in order to optimize the speed and accuracy in evolving a culprit's face.

A second method for increasing the efficiency of the program is to reduce the size of the multidimensional search space. Consider a version of FacePrints that generates female as well as male faces. This requires extending the binary code from 60 to 61 bits. If the additional bit is set to 0, then the face-space is limited to female features and proportions; if it is 1, then male faces will be generated. This reduces the search space by fifty percent, which is clearly advantageous. The witness has the option of selecting the male or female database, then "freezing" the "sex bit" to the appropriate value. Further reductions in the size of the search space can be achieved while a composite is in the process of being generated. For example, if a witness observes that the eyes of a particular composite closely resemble those of the culprit, he or she may "freeze" these eyes by clicking the mouse on those eyes. This operation replaces the eye segments of the binary codes throughout the entire population with the binary sequence that specifies the eyes of the displayed composite. Unless a mutation occurs, all future composites will then have the same eyes as the current composite. Operations such as these greatly enhance the speed of the search process.

Initially such "freezing" operations may seem equivalent to an instructional or Lamarckian-like process, but this is not the case. As noted earlier, the witness is always selecting from options or facial configurations within the range of possibilities stored within or generated by the computer; the witness is not the source of any new variations. This is the essential difference between selection and instruction. What appears to be an instructional process is often just a method for reducing the size of the space that is being searched by a selection procedure.

FacePrints begins with a set of points that are randomly distributed throughout a virtual face-space; selection by the witness causes these points to migrate, over generations, toward the culprit. The efficiency of the migration is greatly enhanced by the sharing of partial solutions through crossover, and by the incessant exploration of variations around these partial solutions, using mutations. The result is not a "random walk" but rather an efficient search process that can find a single face, out of more than a billion billion possibilities, in about two hundred evaluations.[5] Research has revealed that FacePrints can produce useful composites in about sixty percent of cases, compared with a thirty-eight percent success rate using conventional composite systems (Figure 3.3).

What is it about this procedure that makes it work so well? The short answer is that traditional methods are Lamarckian instruction-based systems, while FacePrints involves a Darwinian selection process. The importance of this difference can best be illustrated by thinking about a simpler problem: buying a suit.

SELECTION OR INSTRUCTION?

There are two ways to buy a suit. Most of us visit the local department stores, and after a careful inspection of the available designs and sizes, we select the one we want. I will call this the selection

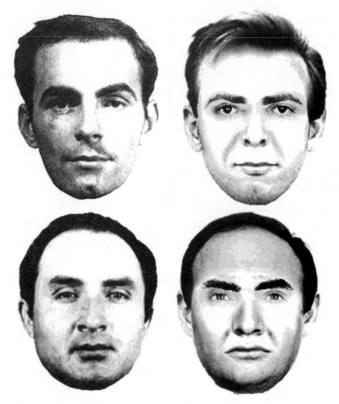

Figure 3.3: Original photographs (left) and composites produced using FacePrints (right)

method. A more affluent or discriminating consumer may prefer a visit to the local tailor. He or she gives the tailor a precise set of instructions regarding the size, style, and material to be used. This is the instructional method. In each case we end up with a suit, but the attributes of the two procedures are quite different.

First of all they differ over the speed of the process. If we are fortunate and the department stores are well stocked, we can arrive home with our new suit on the very same day. In contrast, the tailor may inform us that an appointment for a first fitting might be possible in two weeks, but we should call first! For most of us, however, the real difference comes when we are presented with the bill. The expense of our tailored suit threatens the limit of our gold card, while the cost of our sale-priced item from the department store may be absorbed by eating at home for the next few months. Cost is really a reflection of efficiency. The items in the department store were probably manufactured by highly automated equipment, or low-cost labor, or both. The factory expects to produce and sell many "clones" of your fashionable attire, while the tailor caters to your individual taste and toils, we hope, for a perfect product. This brings us to the question of quality. Here the instructional method is the clear winner. The tailor-made suit is more likely to meet all of our requirements, while the department store item is simply the best compromise that we could find.

But what happens if we modify the problem and consider buying something that is much more complex than a suit—perhaps a car? As we increase the complexity of the item, the balance shifts back toward the selection procedure. Few of us could provide adequate instructions for creating a quality automotive product, and few mechanics could put together a car from our instruction set. In contrast, the showroom car was manufactured in a factory that has made many cars in the past and already possesses all the relevant knowledge. For a selected car we don't have to supply any knowledge at all, since it already exists in the factory. Given these considerations, going to a showroom and selecting from a variety of factory-made designs would be the more prudent choice, even if quality was our major concern.

So far we have concluded that for a complex product, the selection procedure offers a faster and more efficient method for achieving the desired outcome. One final attribute, however, deserves our attention: the question of creative design. We may believe that we have good taste in clothing or cars, but we may also be mistaken. Our choice is limited by our own experience, whereas factories that mass-produce such items, year after year, enjoy the benefits of feedback from many consumers. The factory knows which cars were successful over many previous years, and it can combine the parts of its most successful models to create new designs with a high probability of success. If you think this sounds suspiciously like the race of life discussed in Chapter 1, you are correct. Creative new designs can be achieved by capitalizing on past successes, and the process is cumulative over generations. As a result, the latest Mercedes bears little resemblance to its historical predecessors. Combining the attributes of past successes and adding a light sprinkling of novelty is a powerful prescription for creativity. Compared to its major competitor, the instructional process, selection is a faster and more efficient method for generating complex creative designs. For these very reasons the process appears to underlie a wide variety of adaptive systems, such as biological evolution, the immune system, and even learning and reasoning.[6]

At the end of Chapter 2, I proposed that natural selection, learning, and reasoning can be viewed as three adaptive processes that allow animals to fine-tune their behavior to increasingly rapid environmental change. I now propose that such a biological adaptive system could be modeled as three selection procedures, nested inside one another. The outermost loop, natural selection, would exploit long-term environmental consistencies; the middle loop, learning, would allow modification to faster but consistent change; and the innermost loop, reasoning, would provide a mechanism for adapting to future but predictable variations in the environment. Like three nested Russian dolls, these adaptive processes would possess a similar design based on selection rather than instruction; each would be constrained to some degree by the one outside it; but when working together, they would form a single adaptive unit that has a common destiny. We can learn a great deal about how such a system could be modeled by examining an actual nested adaptive system: the immune system. Once we understand how the immune system learns new responses to cope with rapid changes in the microbial world, we will have a better understanding of how a feeling animal—and ultimately a human with complex emotions—can use the same principles in order to learn new behavioral responses to cope with rapid changes in the physical world.

LEARNING BY SELECTION

A foreign substance that triggers an immune response is known as an antibody generator, or simply as an antigen. When antigens enter the body, they are met by a formidable array of weapons including macrophages and T cells, but eventually they are attacked by antibodies generated within B cells. Even before contact with an antigen, a B cell, during its development, has already generated an enormous variety of antibodies and inserted them into its cell membrane. These antibodies, which are proteins, are pattern-recognizers that fit with varying success onto the surface of a foreign substance. The initial diversity of antibodies, produced by the rearrangement of gene segments within B cells, is so enormous that some will inevitably attach to the antigen. These selected B cells then undergo a phase of rapid cell division, and some eventually differentiate into antibody-secreting plasma cells. During this proliferation phase, however, there is a very high mutation rate in the hypervariable region of the antibody genes within the B cells. Indeed, the rate of such somatic mutations is more than a million times the mutation rate in other cells of the body. These mutations produce many new variations of the already-successful antibodies. As a result of proliferation with somatic mutation, some of the new mutants are inevitably an even better fit to the antigen than the antibodies found on the surface of the original B cells. These cells then increase in number and are the basis of the secondary immune response that protects us from a second infection by the same antigen.

The secondary response has been achieved by a refinement of the original antibodies using a second nested selection procedure. Natural selection, acting over many generations, was responsible for providing the assortment of parts needed for constructing the primary array of antibodies. The refining process, which occurs within each individual's lifetime, depends upon reducing the search space by first selecting the best out of this initial "crude" response, then generating a more "refined" attack by exploring small random variations around this initial solution. That is, the overall immune response depends on a nested system consisting of a "crude" primary response (the outer loop), followed by a "refining" secondary response (the inner loop). The initial set of genes that combine to generate the antibodies of the primary immune response is the product of natural selection—the outer loop. It is the ability of organisms to survive and reproduce that has selected the genes responsible for this adaptive design. This primary response is then refined by the secondary immune response, the inner loop, that occurs within each individual's lifetime.

The "refining mechanism" depends first upon selecting the most successful variants of the primary response and then upon generating new random variations around these "crude" solutions. In this case, however, the definition of fitness is quite different from the fitness measure used by natural selection. Fitness, for the refining mechanism, depends upon the ability of an antibody to attach to an antigen; this affinity selects which B cells reproduce with somatic mutation. It is important to note the contribution of the outer loop to the functional design of the refining mechanism: the laws by which the inner loop learns. Natural selection has provided the secondary immune response with both a proximate value system that defines fitness for the inner loop ("binding is good") and an appropriate response to fitness that increases the probability of generating a more effective adaptation ("proliferate with somatic mutation"). Using this arrangement, the secondary immune system functions in a manner that further enhances gene survival; it is steered in an adaptive direction.

Although they are quite different processes that operate over different time scales, the primary and secondary immune responses are two important examples of adaptive mechanisms that employ selection rather than instruction—natural selection and clonal selection, respectively. In both cases the adaptive response is achieved by selecting from a large array of random variations that preexist in some form of memory. In both cases the role of the environment is not to instruct but merely to select the fittest variations, which are then maintained for future use. Like two nested Russian dolls, these selection processes look alike, but they differ in size and scope. Unlike the primary system, the somatic mutations of the secondary immune response will not be transmitted genetically, since there has been no alteration of the DNA within sperms or ova. The immune system of the infected individual has "learned," using a second selection process. New antibodies are now prepared for any further assault by the offending antigen, but such acquired adaptations are not passed on to future generations; this is not a Lamarckian process. The ability to generate a secondary immune response has evolved by natural selection, but the acquired immunity to a specific offending antigen is a unique solution that is stored only within the body of an individual organism. In these respects the development of a secondary immune response, and behavioral learning, have much in common.

As discussed earlier, the immune system evolved as an answer to the incessant and constantly changing menace of parasitic microorganisms. For more than 300 million years, parasites and the immune systems of multicellular organisms have been locked in a perpetual arms race, in which each side had to counter the innovations of the other or die. For large organisms a major part of the solution was sex. Sexual reproduction generates new variations of genotypes, and the genes of the immune system, the immunoglobulin gene superfamily, exhibit the highest degree of genetic polymorphism. Like the brain, the immune system has evolved a complex integrated organization that is capable of generating and storing adaptive solutions to a wide variety of rapidly changing environmental threats. Also like the brain, it is capable of producing new and effective solutions to novel events that may be unique to individual organisms and may never have been encountered in the history of the species. For these reasons the immune system's ability to learn and store new adaptive responses by making use of two nested selection procedures offers a potential design for an animal that can learn and store new behavioral responses. In other words the design of the secondary immune response provides a model for an adaptive learning mechanism.

An adaptive learning mechanism modeled after the immune system would require an outer loop, designed by natural selection, that would allow animals to adapt to those features of their environment that are stable over generations, and also an inner loop that could modify and refine their behavioral responses to environmental factors that changed within their life span. As is true for the immune system, the inner loop, to be adaptive, would require (1) a proximate value system provided by the outer loop and (2) an appropriate response to this evaluation that would increase the probability of generating an appropriate solution.

As I reflected on these design requirements, it became apparent to me that human feelings were the ideal proximate value system needed for such a learning mechanism. Positive and negative feelings could evolve by natural selection (the outer loop) to reflect the importance of environmental factors that consistently enhanced or decreased the probability of gene survival. Such feelings could then provide the required value system to guide a selection-based learning mechanism (the inner loop).

I soon became intrigued by the possibility of designing a species of computer animal that could "learn" using these principles. Such virtual creatures would have to acquire feelings by natural selection and then use these simulated feelings to guide their learning. I then realized that I could model both of these processes in a computer by using two nested genetic algorithms. The outer loop would acquire feelings by natural selection (phylogenetic learning) and the inner loop would use these feelings to learn from experiences (ontogenetic learning). The result of this Russian-doll design would be virtual animals that could acquire new adaptive responses to cope with the uncertainties of a constantly changing environment.

PROCESSES ARE NOT PROPERTIES

The computer model described in the next chapter introduces some of the functional roles of feelings into the design of an intelligent learning machine. Such models are useful for simulating and evaluating theories, but simulations are just that—simulations. In recent years it has become fashionable to view the human brain as a sort of computer, but like prior metaphors, such as telephone exchanges and holograms, this metaphor has limitations as well as merits. On the one hand, there is no doubt that computer simulations have contributed substantially to our understanding of the processing of information in neural networks; on the other hand, such models have failed to provide an adequate simulation of feelings. While some might claim that computer models offer an ever-improving simulation of thinking, few would make a similar claim for emotions. But computers don't really think, they merely simulate thought, and although they certainly don't feel, they can be used to simulate the functions of feelings.

The simulation of feelings in a digital computer is a modest goal, compared with the aspirations of some cognitive scientists and the practitioners of artificial intelligence (AI). For such scientists the brain is a computer, perhaps a parallel processor, and mental processes, like thinking or feeling, are computational programs that can be executed by any such general-purpose machine. From this perspective there is no reason why future computers should be limited to merely simulating thinking or feeling, because all mental faculties are viewed as computational procedures that are independent of their physical substrate. Such claims are based on the supposition that mental faculties are exclusively algorithmic, and since it is possible to construct a general-purpose computer that can implement any algorithm,[7] it should possible to construct a digital computer that could possesses all such mental faculties. This viewpoint, commonly known as the "strong AI position," has received a great deal of support from a thought experiment that was first posed by the famous computer scientist, A. M. Turing, almost fifty years ago. The "Turing test" presents a dilemma that must be addressed by anyone who claims that computers don't think or never will think.

Imagine, at some time in the distant future, that you are required to discriminate between a computer and a human being. There is a computer in one room and a human being in another. You can't see into the rooms, but your task is to decide which is in which. You can use a keyboard to ask any question you want, and you can examine the answers that are displayed on a screen in front of you. The purpose of this arrangement is to eliminate all physical cues and force you to discriminate on the basis of intellectual capacities alone. What questions could you ask that would allow you to distinguish between the computer and the human, given that the computer has access to lots of information and has been designed to "fool" you? Could you tell the difference with better accuracy, for example, than you can tell the difference between a man and a woman? If not, does this mean that the computer must be thinking? After all, when you speak to another person over the telephone, you automatically assume that they can think on the basis of your questions and their answers. What would it mean if you couldn't tell the difference between a machine and a human in the Turing test?

Not much! Perhaps the most compelling argument against attributing thought processes to a machine that passes the Turing test comes from another thought experiment. Imagine applying the Turing test to a ventriloquist's dummy. Certainly the dummy could answer all your questions in a very human manner. Your conclusion, however, would be, not that the dummy thinks, but simply that the dummy has been programmed by someone who thinks. Isn't this all that can be concluded from passing the Turing test? From this perspective, our propensity to attribute thought processes to other human beings may simply be an accident of our history; we haven't encountered any "cyborgs" in our past. If, however, cyborgs were common, we would certainly be more guarded about attributing human thought processes to others on the basis of questions and answers alone. So what's the difference between human thinking and the algorithmic process that controls the behavior of such cyborgs? The difference, I believe, is that although many human mental processes may be algorithmic in nature, it does not follow that they are solely algorithmic.

An algorithm is simply a list of instructions or a set of rules for carrying out some procedure, such as changing the wheel of a car or multiplying two numbers together. In the real world, for example, we could change a wheel by jacking up a car, unscrewing the lug bolts, removing the wheel, putting on the spare, screwing on the lug bolts, and releasing the jack. This would be a "wheel-change algorithm." Each step of the procedure, however, requires a human operator to supply the meaning for such terms as *lug bolts* and *unscrewing*. Similarly, when a computer multiplies two numbers using an algorithm, it is not computing anything unless a human operator assigns a meaning to the different states of the machine. Meaning is not an inherent property of algorithms, because algorithms describe processes and not the contents of those processes. An algorithm, for example, could possibly describe the process of seeing or feeling but not the qualitative nature of the conscious experience: what is actually seen (redness) or what is consciously felt (pain). In his book *The Rediscovery of the Mind,* John Searle convincingly makes this point, arguing that the molecules in a wall are probably executing a wide variety of algorithms but all such operations are meaningless without an interpretation of these states by a conscious mind.[8] Meaning requires conscious experiences, which are the evolved emergent properties of biological brains.

Of all the different emergent properties that could originate from the organization of neural tissue, only a subset will be selected. Selecting that subset that ensures the survival of DNA that can make brains possessing such emergent properties. In this manner it is always the biologically functional usefulness of conscious experience that dictates and refines the neural organization that is responsible for such attributes. When the functional role of feelings is well understood, there is no reason why such processes could not be simulated in intelligent machines. Processes, however, are not properties. Human conscious experiences are inherent emergent properties that arise from the complex arrangements and interactions between nerve cells, not transistors, so it is highly unlikely that any silicon-based machine could ever generate similar conscious attributes. Nonorganic computers will probably never think or feel in a conscious manner, but there is no reason why these processes cannot be simulated in a variety of different machines. Indeed, a computer simulation of feelings is the next step in our exploration of the evolutionary role of human emotions.

FOUR

❧

Russian Dolls

EVERY MAN, WOMAN, AND CHILD HAS EXPERIENCED a wide variety of emotions, from anger and love to fear and happiness. Statements like "I love you" and "I hate mathematics" and "I am very proud of my daughter" are descriptions of our inner private feelings—emotions—that are not directly observable by other people. In addition to emotions, a second category of inner feelings are those elicited by sensory inputs. These sensory feelings, called *affects,* are directly evoked by specific inputs from the internal or external environment and include such evaluative experiences as hunger and thirst or pain and sweetness, respectively. These affects are feelings whose quality is directly related to the nature of the sensory event that evokes them, and unlike emotions like pride or anger, they occur in the absence of any complex cognitive processes (Figure 4.1). In Chapter 5, I will discuss the nature of such cognitive processes and how they are responsible for a wide variety of different emotions. At this point I want to focus on the shared characteristics of affects and emotions.

The most important characteristic of all feelings—emotions as well as affects—is that they come in two different *hedonic tones,* positive and negative. No feelings are neutral, for the presence of hedonic tone—pleasantness or unpleasantness—defines feelings and distinguishes them from all other types of conscious subjective experiences, like thoughts and sensations. Joy is pleasant and pain is unpleasant; we are never ambivalent about either. *Hedonic tone* refers to the evaluative aspect of a feeling, its pleasantness or unpleasantness, rather than the quality of the experienced sensation (Figure 4.1). Sourness and bitterness are qualitatively different sensations, but they both evoke a negative hedonic tone. Sweetness and sourness are also qualitatively different, but they elicit different hedonic tones, positive and negative, respectively. Redness and blueness are also qualitatively different sensations, but they possess no hedonic tone at all; they are not feelings.

	AFFECTS	EMOTIONS
Quality evoked by *(e.g., pain, anger, joy)*	**Specific sensory inputs**	**Specific production rules**
Hedonic Tone or Direction *(pleasantness or unpleasantness)*	Yes	Yes
Intensity *(i.e., degree of pleasantness or unpleasantness)*	Yes	Yes
Physiological Changes *(e.g., changes in heart rate, blood pressure)*	Yes	Yes
Behaviors *(e.g., crying, smiling)*	Yes	Yes

Figure 4.1: Similarities and differences between the two major types of evaluative feelings: affects and emotions. The major difference is that the quality of an affect (e.g., bitterness) depends upon the nature of a specific internal or external sensory input, whereas the quality of an emotion (e.g., anger) depends upon the activation of a specific production rule.

All feelings, emotions, and affects are an integral part of the most intimate and important aspects of our lives. We strive for happiness, cry with pain, love our children, feel sorrow at the death of a parent, or cherish the pleasures of passionate love. It's remarkable that we attribute so much importance to these inner experiences, given the fact that they are not real objects in the physical world, like hearts and kidneys. They don't appear to be critical for life. We can't hold feelings in our hands, look at them, hear them, or directly measure them. They exist only as fuzzy ephemeral internal subjective experiences that flit through our minds. Yet despite their volatile and transitory nature, we treasure our feelings, and they appear to play a central role in our lives. They add texture to our existence, and without them our lives would be meaningless and robotic. We may think it strange that David was haunted and paralyzed by fear of his imagined monsters, but we don't view our own feelings as "strange ghosts" that haunt and control our behavior.

Those same aliens that were curious about our dreams would certainly be greatly puzzled by human feelings. They might observe, for example, that certain patterns of electromagnetic stimulation increase the heart rate and blood pressure of their human subjects. Alien dissertations might be written concerning the fact that humans are a sexually reproducing species with "strange, inner, ghostlike, states" that can be elicited by showing them nude pictures of the opposite sex. To us, however, these inner feelings are just part of the way we are. We don't question them, any more than we question why water comes out of our eyes when we are sad, or why we have eyebrows. We accept our feelings as part of our natural design, and we seldom question how they came to be or what function they play in our lives.

The first insight that is necessary for understanding the origin and function of human feelings is the realization that they are not learned. This statement does not imply that the events that elicit a particular feeling do not become modified as a function of experience; they certainly do. What I am saying is that the feelings themselves remain constant despite considerable changes in the eliciting events. Indeed, it is difficult to imagine how someone could learn a completely new internal feeling as a result of an experience or instruction of any kind. The situation would be analogous to an animal learning to grow its hair when the weather gets cold. Growing hair entails a complex set of internal cellular processes that involve the transcription of messages from DNA and their subsequent translation into specific proteins. The environmental event—a drop in temperature—is simply a decrease in the average kinetic energy of random air molecules striking the skin. There is simply no way that a change in a random environmental activity can teach an animal the complex cellular events that are required to grow hair. There is clearly insufficient information in the signal to provide the instructions that are needed to regulate the intricate hair-growth process. Natural selection appears to offer the only possible explanation for how such an adaptation could arise.

In an analogous manner it is not possible for a simple environmental input, such as a sugar molecule, to teach an organism how to feel a conscious subjective feeling, like sweetness. Nor is it any more plausible to suppose that another human being, rather than the environmental signal, is responsible for imparting this instruction. Imagine trying to teach a child that it should feel anger when spanked. The child asks, "Exactly what should I feel when you spank me?" There doesn't appear to be any way to explain what anger feels like if it is not among the repertoire of feelings that the child already possesses. Furthermore, the instructions must also include "You should not feel anger when I give you sugar!" That is, an effective teaching process would not only have to teach an individual what a feeling is like, it would also have to specify the unique set of circumstances under which the feeling should be generated. It is obvious that neither the environment nor any another human being could perform this task. Feelings, however, as emergent properties of the nervous system, can evolve by natural selection.

The second important observation about feelings is that their pleasantness or unpleasantness is closely related to whether the events that evoke them are, or were, likely to enhance or decrease the survival of our genes. For example, we don't feel happy when our child dies, and we don't grieve over a sudden increase in our resources. These two circumstances elicit qualitatively different feelings with opposite hedonic tones—negative and positive, respectively. In Chapter 5 we will examine the full range of human emotions and discuss how each one is related to a specific aspect of our reproductive success. For now, it is important only to observe that the relationship between hedonic tone and gene survival is also true for our sensory feelings. Tissue damage evokes the negative affect of pain, a potential threat to our survival, whereas sugars have a pleasant sweet taste and were biologically favorable throughout our long ancestral history. A contaminated food, like spoiled milk, evokes a sour taste or noxious odor, and it is certainly no accident that our most intense positive feelings are associated with orgasms, when we are closest to perpetuating our genes.

Indeed, this observed relationship between hedonic tone and gene survival is the strongest argument against the notion that conscious experiences are irrelevant epiphenomena. If conscious feelings were irrelevant, there would be no reason why this relationship should exist. Although this is an interesting observation, however, it doesn't explain the origin or function of feelings. A comprehensive theory of feelings should explain how this relationship is established, how it is maintained as new events acquire the ability to elicit feelings during our lifetime, and the specific functional role of the many qualitatively different feelings that we possess.

Affects are the simpler of the two kinds of evaluative feelings, so we will examine them first. Despite their simplicity, however, affects possess hedonic tone and intensity, attributes that they share with even the most sophisticated emotions (Figure 4.1). Understanding the evolution and functional role of affects will provide insights into the origin and function of our more complex emotional states. Together, affects and emotions will prove to be integral to our survival—not the irrelevant epiphenomena that many cognitive scientists believe them to be.

THE IMPORTANCE OF SENSORY EVALUATIONS

All animals evaluate their environment with or without the benefit of feelings. Even a unicellular organism, like an amoeba, accepts or rejects objects that come into contact with the surface of its body. A more complex arrangement is found in the paramecium, another single-cell animal. Here small particles, mostly bacteria, are taken into the body through a specific entrance, an oral groove. The food vacuoles containing bacteria and a minute drop of water then circulate through the paramecium's body, where they are subjected to digestive processes. Finally the waste products exit through an anal pore. Interestingly, more food vacuoles are formed for certain kinds of bacteria than for others, so the paramecium appears, however imperfectly, to discriminate. Also, unlike an

amoeba, whose entire body surface can absorb nutrients, a paramecium has distinct zones for ingestion and excretion; other areas
of its body are covered with cilia, which have different sensory or
motor functions. When cilia are touched with a large object such
as a glass rod, a paramecium exhibits a "phobic response." This
involves a rapid reversal in direction, followed by a random change
in the orientation of its body axis. The primitive ability to engulf
what is useful for biological survival (nutrient) and reject or escape
from potential danger (toxins or tissue damage) is present in even
the simplest single-cell animals. Such evaluative discriminations
continue to be fundamentally important for the survival of multicellular organisms, which must swallow and digest foods but spit,
vomit, sneeze, or cough back into the environment anything that
is potentially dangerous. The inherent problems involved in
approaching, recognizing, and ingesting food while avoiding toxins and other potential threats appear to have provoked the evolution of the first evaluative sensory feelings.[1]

In complex organisms like humans, highly evolved taste and
olfactory organs guard the entrances to the digestive and respira-
tory systems. These sophisticated chemical detectors let clean air
and pure water pass but respond to nutrients or contaminants with
a barrage of neural inputs that evoke inner feelings of either pleasure or distaste. Ubiquitous sensors throughout the body translate
tissue damage, or excessive temperature, into evaluative sensations
like pain or discomfort, while a change in the blood sugar level or
the salt/water ratio, detected by internal sensors, can elicit such
negative affects as hunger or thirst. These basic sensory feelings are
evoked by relatively simple internal or external stimuli that have
been commonly present and repeatedly encountered in ancestral
environments. Consistent environments over many generations
provide an opportunity for such feelings to evolve by natural selection. Indeed, a human child clearly doesn't need to learn to like the
taste of sugar or to feel pain when pinched. Such evaluative feelings depend only on the nature of the stimulus and the internal
state of the body. The pleasurable taste of sugar, or the unpleasantness of pain, may be influenced by internal factors, such as blood
sugar or endorphin levels, but such sensory feelings occur in the
absence of any learning. The feelings required for surviving and

maintaining the integrity of the body are part of human nature, and they can provide the necessary value system for a selection-based learning mechanism, modeled after the secondary immune response discussed in Chapter 3.

If an animal is bitten, stung, or becomes ill after approaching or ingesting another animal or plant, it would clearly be advantageous for it to inhibit or avoid such behaviors on future occasions. Similarly, it would be beneficial for the animal to retain and repeat behaviors that have led to new sources of nutrition or have provided some protection against the elements. Such flexible behavior-control mechanisms would be a major advance over the rigid approach-and-avoidance behaviors observed in less complex organisms. The inflexible evaluative discriminations of the paramecium may be useful in a constant environment, but such rigid adaptations offer little advantage or protection to organisms that are faced with environments that harbor rapidly changing benefits or threats to their survival.

Two simple hypotheses arise from these considerations. The first is that sensory feelings evolved in response to those environmental events that have consistently presented opportunities or threats to biological survival in ancestral environments; and second, behaviors followed by positive feelings are facilitated, whereas behaviors followed by negative feelings are inhibited. These two hypotheses have been evaluated by a computer model that was designed to simulate both the acquisition of affects by natural selection (an "outer" genetic algorithm), and the use of such simulated feelings by a learning mechanism (an "inner" genetic algorithm). This is the Russian-dolls design described in Chapter 3. If the simulation is successful, then the outer loop, shaped by natural selection, should be capable of evolving appropriate simulated feelings that allow a species of computer animals to learn from their experiences.

EVOLVING SENSORY EVALUATIONS

Imagine a species of computer animal (Sniffers) that, not unlike some dogs, survive by following the odor trails left by their prey. In this simulation it is proposed that each Sniffer has a nose that can detect odors, and the strength of an odor will be greatest when a Sniffer is in close proximity to a trail. That is, the strength of an odor (S) increases as a Sniffer gets closer to an odor trail (Figure 4.2). For simplicity it is assumed that the ability to catch prey is the single factor that governs Sniffers' survival. Accurate Sniffers, who can closely follow odor trails, are assumed to catch more prey and have higher survival rates than inaccurate Sniffers. We will also assume that over its lifetime each Sniffer has an opportunity to follow ten different odor trails, with fifty learning trials per trail, so the probability of survival of any individual Sniffer can be estimated from its average accuracy over all five hundred attempts. This fitness measure is used by the outer genetic algorithm, simulating evolution, to determine which Sniffers survive and reproduce in each generation. Stated simply, accurate Sniffers—those who can rapidly learn to track odor trails—have a higher probability of surviving and reproducing than less accurate Sniffers.

An outer genetic algorithm evolves the structure of Sniffers'

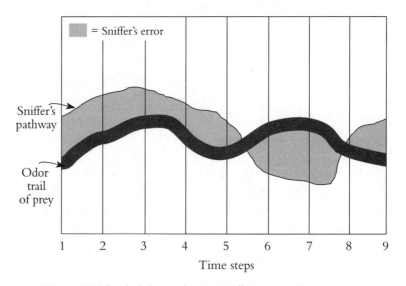

Figure 4.2: The shaded area shows a Sniffer's error as it moves across the computer screen in nine time steps. The strength of an odor (S) is a function of a Sniffer's proximity to the trail. The proximity is measured by the total area minus the area of error.

feelings.[2] At the beginning of the simulation, environmental odors have no fixed pleasantness or unpleasantness for Sniffers. For any individual Sniffer the simulated pleasantness or unpleasantness of an odor is governed by the value of two inherited genes, H and a. The reason for using two genes to specify Sniffers' feelings is to permit such simulated feelings to vary over a wide range of positive and negative values. The H gene (hedonic tone) specifies whether a feeling is positive, negative, or neutral; it can evolve to a value of $+1$, -1, or 0. The a gene is an amplification factor that will determine the intensity of the positive or negative feeling; it can evolve to any value between 0 and 7. Over the course of the simulation, Sniffers' simulated feelings will evolve toward specific values for H and a. To examine the evolution of these simulated feelings, I initially set the H and a genes to random values in the first generation of Sniffers and was interested in finding the final values of these genes after many generations.

Within each Sniffer the values of H and a are coded on a short binary string. For example, 11 010 would be decoded as $+1$ and 2, for the values of H and a respectively; 01 111 would be interpreted as $H = -1$, and $a = 7$. (For the H gene, the first bit specifies its sign and the second bit specifies its value.) These two genes specify the positive or negative hedonic tone (H) and the amplification (a) of the simulated affect that a particular Sniffer experiences when it is close to an odor trail.

$$\text{Magnitude of Affect} = (H) \times (S)^a$$

The magnitude of an affect (degree of pleasantness or unpleasantness) is equal to the hedonic tone (H) multiplied by the strength of the odor (S) raised to the power a, the amplification factor. If the H gene is 0, for example, then a Sniffer experiences no affect when it is close to an odor trail. If H is $+1$, then it experiences a positive affect. A -1 value for H would be interpreted as an unpleasant smell. That is, the H gene determines the hedonic tone (positive or negative) associated with any odor, while the a gene determines the amplification of the simulated affect. If a is equal to 1, then the magnitude of a simulated affect is a linear function of the intensity of the smell. A large value for the a gene would result in a very large positive or negative affect when a Sniffer is close to a trail.

Both the *H* and *a* genes are set to random values in the first generation of thirty Sniffers. These affect genes are the binary strings that cross over and mutate between generations. That is, each member of a first generation of thirty Sniffers begins its life with a short random binary string, its affect genes. Each Sniffer is then presented with the problem of learning how to follow ten different odor trails, with fifty learning trials per trail. For each of these first-generation Sniffers, the average accuracy in following these trails, the fitness measure, is computed. Sniffers breed in proportion to this measure of their fitness. During breeding the standard genetic algorithm described in Chapter 3 is used to cross over and mutate the affect genes between breeding pairs. Since the fittest animals breed more than the less fit animals, their genes, or some variation of them, will be disproportionately represented in the next generation. This is the design of the outer genetic algorithm that simulates natural selection.

Each Sniffer, however, also possesses an inner genetic algorithm, the learning algorithm, that permits it to learn how to follow trails within its own lifetime.

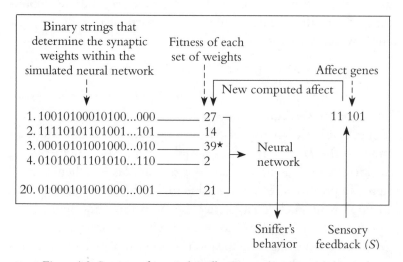

Figure 4.3: Structure of a typical Sniffer. On any learning trial the network weights are derived by crossing over the current fittest weights (★) and a second set of weights selected in proportion to fitness.

In humans, learning depends on the efficiency of communication between nerve cells; so too with Sniffers. Each Sniffer possesses a simulated neural network (unique to each Sniffer) that permits it to move across the computer screen in nine time steps (see Figure 4.2). Its position at any time step is governed by the synaptic weights within its neural network. In a neural network model, these weights are used to simulate the efficiency of the synaptic junctions between nerve cells. On the first trial of any learning problem, the weights, which are coded on long binary strings, are set to random values.

When learning to follow a new trail, the Sniffer's initial movements across the screen will be random since they are determined by weights supplied from the initial set of random binary strings (see Figure 4.3). Soon, however, these weights will be adjusted by the inner genetic algorithm (the learning algorithm), which employs the magnitude of positive or negative affect as a measure of the fitness associated with any set of weights. That is, the genetic algorithm within each Sniffer searches for the particular set of weights necessary to follow any specific odor trail by using the magnitude of the simulated affect that the animal experiences, on any trial, as a measure of the fitness of the set of weights that were used on that trial. Suppose, for example, that a particular Sniffer by chance possesses a good set of synaptic weights, so that on one particular trial it somewhat closely tracks a certain odor trail and, at the same time, its affect genes dictate a positive affect. The magnitude of the experienced affect would be computed as a high positive value. This high positive value would be treated as a high fitness value for the specific set of weights used on that trial, when they are being adjusted by the inner genetic algorithm that simulates learning. As a result the inner algorithm would, over a set of learning trials, rapidly develop an even better set of synaptic weights for tracking that particular trail. If, during its lifetime, the affect genes of this particular Sniffer also facilitated its ability to learn how to follow many different odor trails, then this animal would catch more prey, and it would have a higher probability of passing on its affect genes to future generations. In essence, good learners survive and reproduce better than poor learners, so every new generation of Sniffers is better at learning than the previous generation. Sniffers don't transmit what they have learned to future generations; they pass on only the values of the affect genes, *H* and *a,* that permit them to be such efficient learners. This is not a Lamarckian "inheritance of acquired characteristics." In this model both learning and evolution are implemented as Darwinian selection procedures.

In summary, the behavior of Sniffers is controlled by two nested genetic algorithms. The binary strings of the outer loop provide the values of the two genes (*H* and *a*) that control the magnitude of the pleasantness or unpleasantness that a Sniffer can experience when it is in close proximity to an odor trail; the binary strings of the inner loop are interpreted as sets of weights within a simulated neural network that governs Sniffers' movements. Over learning trials the inner (learning) algorithm adjusts the weights within the neural network, using the degree of pleasantness or unpleasantness experienced on any trial as evaluative feedback. The outer algorithm, simulating evolution, adjusts the hedonic tone and amplification parameters of the affect genes using the behavioral ability to catch prey (trail-following accuracy over a lifetime) as the fitness measure. At the start of the simulation, all binary strings, both the synaptic weights and the affect genes, are set to random values in a first generation of thirty Sniffers. The simulation then begins.

The first generation of Sniffers performs poorly as they wander across the computer screen with little or no regard for the location of an odor trail. Some, however, by virtue of possessing a fortuitously good affective value system, improve their performance over the series of trials on any trail. That is, when these Sniffers are confronted with an odor trail, the learning algorithm rapidly adjusts the weights within their simulated network, using their fortuitously good affective value system to guide the learning process. Many less fortunate Sniffers learn to stay far away from odor trails, and many others learn nothing at all. At the end of the first generation, however, it is the good learners who have caught the most prey, and they alone pass on (with crossover and mutation) their affect genes to the next generation. Over generations Sniffers rapidly evolve a highly competent affective value system that permits them to quickly learn how to track any odor trail that they encounter during their lifetime. This affective value system quickly spreads throughout the entire population.

REVEALING OUTCOMES

The Sniffer simulation reveals a number of interesting attributes that are important for the performance of an intelligent learning mechanism. First it demonstrates that a simulated neural network can be designed to learn by selection rather than by instruction. Beginning with random synaptic weights, simulated feelings by themselves are sufficient to find the appropriate set of weights needed to master this simple task. Each Sniffer's learning mechanism can be viewed as a hypothesis-testing procedure. Hypotheses are generated by the random variations of synaptic weights around their current values, while feedback from the affective value system dictates which of the many generated hypotheses are retained and stored as permanent modifications to the nervous system. Learning in this model is a discovery process, and the more pleasant the discovery, the better it is retained.

For organisms using such a design, the hypothesis-generating procedure is not totally deterministic. Although at any time the synaptic connections within the nervous system are a product of the organism's genetics (initial weights) and past experiences (modified weights), constructing each new hypothesis involves a random exploration centered around the current values. Randomness, rather than being a destructive process, provides a unique mechanism for exploring new creative variations of previously stored solutions. In both Darwinian biological evolution and selection-based learning, the random element serves as an effective procedure for exploring a multitude of new and sometimes creative possibilities centered on past success. Random variations of genes or synaptic weights provide a method for discovering new creative solutions without the limitations that some preconceived direction of exploration would impose. The probabilistic nature of these hypothesis-generating procedures, however, precludes a totally deterministic description of exactly how or when a creative insight will occur.

A second insight arising from the Sniffer simulation concerns the nature of the proximate value system that evolves as a consequence of natural selection. As the simulation reveals, the hedonic tone of the simulated affect evolved over generations to reflect the value of the environmental factor that was important for gene survival. Now consider how Sniffers would interact with a second odor trail—perhaps that of a predator that eats Sniffers for lunch. How would the affect genes evolve under these conditions? Using the same formula to compute simulated affect, a number of first-generation Sniffers would inevitably experience a positive affect when close to such trails and would quickly learn to follow them. Such predator-following Sniffers, however, would put themselves in grave danger and would undoubtedly run a higher risk of death or severe injury. They would have a short life, and very few would pass on their affect genes to future generations. Clearly odor trails left by predators could never come to reliably evoke a positive affect.

Some more fortunate first-generation Sniffers, however, would inevitably experience a large negative affect when they followed a predator's trail. This high degree of unpleasantness would assign a very low fitness value to the synaptic weights responsible for following such odor trails. As a consequence, this predator-following behavior would soon disappear, as these Sniffers learned to avoid such dangerous trails. Such Sniffers would survive and pass on their negative affective structure to future generations. Soon all Sniffers would find the trails of predators to have a very unpleasant smell and would consistently avoid them. If this is the case in living as well as simulated organisms, then we should expect positive affects to be elicited by events that enhance biological survival and negative affects to reflect potential threats to survival and reproduction. Indeed, for humans, such relationships are so consistent and reliable that we often think of pleasant or unpleasant affects, such as the sweetness or bitterness of foods, as properties of chemicals themselves rather than as evolved properties of our nervous system. But as discussed in Chapter 1, molecules are just molecules, nothing more and nothing less. Sensory feelings, however, are emergent properties of a neural organization that can evolve, over generations, to reflect the importance of biologically relevant environmental factors.

What would happen if there were other odor trails that had no direct reproductive consequences? Learning to follow such trails would be a waste of time, so Sniffers experiencing either a positive or negative affect in the presence of such odors would inevitably be distracted from tasks that have real biological value. As a consequence, the affect parameter H will evolve to 0 over generations. That is, Sniffers will evolve no hedonic tone to environmental odors that have no biological significance. All of these outcomes—positive, negative, or neutral hedonic tone—are evident in real biological organisms. Sugar now elicits a sweet taste that is pleasant; waste products have distinctive tastes and odors that are regarded as unpleasant; and a host of environmental events that have no direct biological value elicit no hedonic tone at all.

In the Sniffer simulation S could have been any sensory event in the environment, so that what was true for odors would also be true for any event that could be detected in any sensory modality. For example, a Sniffer could evolve a negative hedonic tone—pain—to a tactile stimulation that reduced its reproductive success. Animals designed like Sniffers will always evolve positive affects to events that are predictive of enhanced reproductive success, and negative affects will be evoked by events that predict a decrease in biological survival. Since this affective value system defines the conditions under which learning occurs, such learning will inevitably be steered in an adaptive direction.

Learning, from this viewpoint, is not a general-purpose mechanism that allows all environmental relationships to be acquired with equal proficiency. Instead, it is a constrained mechanism that depends upon an affective value system that provides an immediate appraisal of only those events that have important reproductive consequences. Life and death serves as the value system for directing biological evolution (the outer genetic algorithm), but it is the "omens" of life and death, positive or negative feelings, that direct the learning process (the inner genetic algorithm).

In the Sniffer simulation the fate of the amplification gene (a), regulating the magnitude of hedonic tone, is equally revealing: over generations it always evolved to its maximum possible value. In the simulation, the parameter a could have acquired any value between 0 and 7 (binary 000 to 111). If a had evolved to 1, then the simulated affect would be a linear function of the strength of the odor trail. Over multiple simulations, however, the parameter a did not converge on one; it always evolved to its maximum value of 7. It appears, therefore, that the selection-based learning mechanism is most efficient when the magnitude of the simulated affect is a power, rather than linear, function of the environmental input.

The intensity of the Sniffers' feelings, however, is not simply an amplification of the sensory input (S); it is an amplification of the reproductive consequences associated with the sensory input. In Sniffers only events that were important for gene survival acquired the ability to elicit a positive or negative hedonic tone, and this valence always evolved to reflect accurately the future biological consequences of those events. In this manner positive or negative simulated feelings became the "omens" of future reproductive success. But to provide the instantaneous feedback required for efficient learning, the amplitude of the hedonic tone increased over generations until it became an exaggerated predictor of future reproductive consequences. That is, affects were more functionally useful when they evolved to amplify, rather than accurately represent, the biologically relevant events in the Sniffers' world.

This amplification is clearly evident in real animals. In biological organisms the tissue damage and reproductive consequences of a pinprick are, on average, very small. The immediate pain, however, is an evolved conscious experience that exaggerates the biological threat, because such instant amplification is required for learning to avoid such events. If this design underlies all human feelings, then we should expect them to be exquisitely sensitive to any environmental variables that have, or had, any direct impact on reproductive success in our ancestral environments. The relationship between human feelings, reproductive success, and learning will be further examined in Chapter 5.

The Sniffer model was designed to show that learning could be achieved by a mechanism that depends on selection, rather than instruction. In this model, the genetic algorithm that simulated learning explored variations in synaptic weights (synaptic "mutations") and then selected and retained any such weights that were followed by a simulated positive affect. Using this simple procedure, Sniffers were able to acquire and store the set of synaptic weights required to follow any odor trail. But how could such a design be implemented in the brains of biological organisms?

In real brains, the Sniffer model can be implemented by two global arousal systems, each controlled by an organism's feelings. The first arousal system is responsible for generating small temporary modifications to synaptic weights around their current values—this is equivalent to Sniffers' synaptic "mutations." When this arousal system is active, the organism explores variations of its stored hypotheses. The degree of arousal (and therefore the degree of exploration) is regulated by feelings associated with negative hedonic tone, or the loss of positive hedonic tone. In a simple organism, for example, the affects of hunger or thirst would activate this arousal system. As we will see later, this hypothesis-generating arousal can also be initiated by emotional states. For now, however, it is important only to note that this first arousal system explores small random variations of stored knowledge by generating temporary modifications to the synaptic connections between nerve cells.

The second arousal system provides the necessary feedback for evaluating the hypotheses generated by the first system. In a simple organism, for example, a novel behavior may be rewarded with food. These feelings of reward activate the second arousal system that will retain the synaptic weights that led to that behavior. Unlike the hypothesis-generating arousal mechanism, it is an increase in positive hedonic tone or a decrease in negative hedonic tone that regulates the degree of arousal in this second system. Working together, these two global arousal systems are all that is required for a mechanism that can generate and evaluate hypotheses—a selection-based learning mechanism.

One feature of such a learning mechanism is its nondeterministic nature. In this model of learning, random variations of previously stored solutions generate new creative hypotheses. As a consequence, it is not possible to predict exactly where or when a new creative insight will occur. The model does predict, however, that creative insights are much more likely to occur in individuals who possess a large amount of prior stored knowledge when they are in a state of high emotional arousal, and this appears to be the case.

It is not a cruel accident that some of the most creative and influential writers, artists, and composers have been afflicted with the major mood disorder known as manic-depressive illness. Vincent van Gogh, William Blake, Lord Byron, Tennyson, Poe, and Schumann all suffered the swings in mood, from mania to depression, that characterize this disorder. Many studies have shown that the incidence of manic-depressive disorder is, on average, about six times higher in creative artists, writers, and poets than in the general population. There is also strong evidence for the heritability of this disorder. When one identical twin is manic-depressive, the other twin has more than a seventy percent chance of having the disorder, while a fraternal twin has only a twenty percent chance. In its extreme form, hypermania is associated with impulsive behavior, disjointed thoughts, and poor judgment. But the diagnostic criteria for the milder state of mania, known as hypomania, include "sharpened and unusually creative thinking and increased productivity." Indeed, many afflicted individuals are reluctant to seek a "cure" for their disorder. Unfortunately, when left untreated, the illness often worsens over time, and suicide is a common outcome. The exact biochemical basis of manic-depression is still unknown, but it responds well to treatment with the simple element lithium.

The observation that a specific chemical disorder underlies the creative process in so many different domains of human endeavor is strong evidence for the existence of a global mechanism influencing the creative process. Of course, each of the afflicted artists had their unique domain-specific expertise, developed over many years of hard work and extensive practice of their skills. Nevertheless, the bursts of creativity associated with hypomania suggest that periods of augmented emotional arousal are conducive to new creative insights within each artist's specific domain. It is during these phases of unrestrained emotional arousal that such artists are most productive and creative. In the next chapter we will see that even in ordinary people like you and me, emotions play a central role in regulating the creative processes of learning and reasoning.

FIVE

❧

The Omens of Fitness

IMAGINE YOU ARE SITTING IN AN AUDITORIUM LIS-
tening to a public lecture. Unexpectedly the person next to you,
an attractive young woman you have never seen before, reaches out
and gently touches the back of your hand; nothing else. What
would it feel like? Undoubtedly, like most human beings, you
would experience a sudden surge of emotion. Innumerable questions might flit through your mind, as the topic of the lecture fades
into the background. What did that mean? Why did she do that?
Some individuals might experience anger, others excitement, but
almost no one would be totally indifferent to the event.

What's happening in this situation? Why is our brain responding in such an outrageous manner to such a trivial sensory input?
Our own clothes, after all, are generating similar sensory inputs all
the time, but we are seldom even conscious of these events. The
sensory event by itself cannot be responsible for generating the
feelings that hijacked our conscious awareness and triggered off our
long sequence of questions. Perhaps it's the unexpectedness.
Certainly, as we will discuss later, the unexpectedness contributes
to the intensity of the emotion, but much more is involved. After

all, an unexpected growl would evoke a completely different feeling—fear.

Consider a second situation, in which the emotional reaction is more predictable and probably more intense. You come home from work, and your spouse casually announces that he or she has "found a new lover." Once again these four simple words may evoke an enormous emotional reaction that could persist for days, weeks, or many months, and during this time you may have great difficulty thinking about anything else. Few of us would be "filled with joy" over our spouse's good fortune! Anger is the common emotion, and whether the announcement was expected or not, it would certainly feel very unpleasant for most of us.

Emotions differ from affects in that they possess distinct qualities that are not a function of sensory inputs. Different social situations evoke different emotions, but they all belong to a relatively small set of emotions that we humans possess.[1] If, at a later date, we were telling a friend about the incidents described above, we might say "I was very excited" or "I was very angry." We can give names to our emotions, and we expect our friends to know exactly what each one feels like. For such a large number of different social circumstances to evoke a much smaller set of universal emotions, there must be some general production rules, or contingencies, that produce the same feelings in most human beings.

Furthermore, these contingencies must involve general attributes that are common to many different social situations. For example, we could propose that an emotion like fear is evoked by the expectation of events that elicit a negative hedonic tone. This statement defines a very general production rule that is not situation-specific but that applies to the expectation of any number of unpleasant events, such as a pending surgical operation or the approach of an ominous stranger. Such general production rules could evolve by natural selection if they offered adaptive solutions to a variety of circumstances that were commonly present and repeatedly encountered in ancestral environments. But what is it about emotions that make them adaptive? Or more specifically, how is the survival of our genes related to the many different emotions that we commonly experience?

Emotions are difficult to study since they cannot be measured directly and we are forced to rely on verbal reports. The problem is more complicated in very young children, when even a verbal report may not be available. Fortunately emotions are often accompanied by physiological changes, such as variations in heart rate and blood pressure, and a variety of behaviors, like laughing and crying. Also, specific facial expressions, such as smiling and frowning, appear to be closely related to our inner feelings (Figure 4.1). As a result we can often infer that a specific emotional state is present by observing the physiological or behavioral changes that are reliably associated with our inner feelings. When we are frightened, our heart beats stronger and faster, as blood is diverted from the digestive system into our skeletal musculature. These physiological changes are clearly useful if we are preparing for fight or flight. Facial expressions and other behaviors can also be viewed as useful adaptations, since they appear to influence the behavior of other people, often with beneficial consequences.

It is much more difficult, however, to propose a functional role for our inner conscious feelings, such as envy or joy—all of which have remarkably different qualities. Unlike the physiological or behavioral responses that have an apparent biological importance, the inner feelings themselves don't appear to be a necessary component of emotions. Do we really need these inner subjective experiences? Couldn't an intelligent being, perhaps like Mr. Spock in *Star Trek,* survive perfectly well without them? For humans these inner conscious experiences appear to be important, if only because they figure so prominently in our lives, but if they are more than irrelevant epiphenomena, then we need to understand their behavioral consequences and how these consequences could contribute to biological survival. What is the functional importance of our internal conscious feelings?

THE QUALITY OF FEELINGS

In Chapter 3, I proposed that natural selection could evolve a value system that individuals could use to learn from their experiences. I suggested that the advantage of learning lay in its ability to permit individuals to adapt to rapidly changing aspects of their environment, an advantage that could not be achieved by natural selection alone. Such an evolved value system could specify the circumstances under which learning occurs and thus ensure that learning was steered in an adaptive direction. That is, learning could complement biological evolution by allowing individuals to discover how to better survive and reproduce within their own unique and changing environment. In Chapter 4, I introduced a computer simulation (Sniffer) that used two nested selection procedures to simulate natural selection and learning. In this Russian-dolls design the outer genetic algorithm simulated natural selection and evolved a value system that the inner genetic algorithm then used to learn from its environmental experiences. You'll recall that we found that (1) over generations the simulated positive or negative affects inevitably evolved to reflect the biological importance of sensory events, and (2) such affects were more functionally useful in a learning paradigm when they amplified, rather than accurately reflected, the biological consequences of these environmental events. What I am proposing, then, is that the real importance of feelings—emotions as well as affects—lies in the role they play in regulating how, what, and when we learn and in determining how we reason about the world around us.

While the Sniffer program was useful, it simulated only bodily affects—the magnitude of the hedonic tone associated with the inputs that could be detected by Sniffers' senses. This alone was sufficient for Sniffers to learn. Emotional feelings, however, have an additional component: each possesses its own distinct quality that is not of sensory origin. Anger does not feel the same as fear, and pride is different from happiness. Anger and fear may both be experienced as intense negative feelings, but they certainly feel different; they have different qualities. If hedonic tone is sufficient for learning to occur, then why have we evolved so many qualitatively different inner feelings?

Would there be some additional advantage, for example, for Sniffers to evolve qualitatively different positive feelings to the "smell of food" and the "smell of a mate"? Both events are biologically important, but they are concerned with different domains of reproductive success (survival and reproduction, respectively), and behaviors learned in one context could be very inappropriate in the other context! In these different situations would there be a benefit to experiencing qualitatively different feelings, feelings that discriminated between different kinds of threats or benefits to reproductive success? To understand the function of our many different feelings, it is necessary to delve into the different methods by which humans can enhance their reproductive success.

The fundamental requirements for reproductive success are the ability to survive to reproductive age, to reproduce, and to ensure that any offspring also reach reproductive age. Any property of an individual that increases the effectiveness of any or all of these factors will increase the probability that that individual's genes will be represented in future generations. Qualitatively different feelings appear to be associated with these different aspects of reproductive success. Fear, for example, may be elicited by the expectation of events that threaten personal survival. Sadness may be experienced following the loss of a person, such as a lover, who was associated with a great deal of positive affect; such circumstances forebode a marked decline in reproductive success. In general I am proposing that qualitatively different feelings arise from environmental contingencies that foreshadow fluctuations in the different aspects of reproductive success, while their hedonic tone—pleasantness or unpleasantness—reflects whether such changes represent a net gain or loss. In this manner our feelings act like active filters, or what I call *discriminant hedonic amplifiers,* that define and exaggerate the reproductive consequences of environmental or social events associated with relatively minor fluctuations in reproductive potential. Each qualitatively different feeling appears to monitor a different aspect of reproductive success.

ALTRUISM

Humans may increase their odds of reproductive success in several ways. One way is to help others. At first altruism appears to make no biological sense, since contributing to the survival or reproduction of another human being would certainly decrease one's own reproductive success. That is, the survival odds of the other individual's genes would be increased at a cost to the genes of the altruist. This effect is compounded when one considers that reproductive success is always relative to the reproductive success of others. Indeed biologists use a special term, *fitness,* to refer to this relative reproductive success. Having ten children may appear to indicate high reproductive success, but it would represent low fitness if everyone else were having twenty children. Given these considerations, it would appear that altruism could never evolve: it decreases fitness. That is, if altruists bestowed their kindness indiscriminately, they would soon disappear, and their less benevolent neighbors, who enjoyed the benefits of the altruism without paying the costs, would soon dominate. Under some circumstances, however, altruism can be very adaptive indeed.

Imagine that Ashley has an abundance of resources and Justin is in need. Now, if Ashley helps Justin when the cost to Ashley is small and the benefit to Justin is large, and Justin repays Ashley when the conditions are reversed, then both individuals have enhanced their reproductive success through the use of reciprocal altruism. Over a lifetime individuals who practice reciprocal altruism, or "fairness," would survive very well indeed, and unlike the nondiscriminating benefactors discussed above, they would have a higher probability of surviving and passing on their genes to future generations.

But receiving a gift without reciprocating is also an adaptive strategy, at least for the cheat. To be a viable mechanism for mutual survival, reciprocal altruism must be carefully monitored, since it is fraught with risks for both parties. Feelings such as guilt (when we fail to repay a favor) and anger (when we are not repaid) are obvious candidates for monitoring such transactions. Such feelings can provide an estimate of the magnitude of potential reproductive costs and benefits at the time when individuals are actively engaged in such social transactions, before the actual costs and benefits to reproduction have been realized. Of course we don't think at all about reproductive success when we interact with our friends; nor do we have to—our feelings do the "thinking" for us!

It is worth examining the requirements for reciprocal altruism in more detail, since they shed some light on how feelings would be designed if they were to serve as a barometer for reproductive success. First of all, the practice of reciprocal altruism requires a long-term memory, the ability to quantify resources, and the ability to recognize and discriminate among individuals.

These prerequisites are not fully developed in very young children, so reciprocal altruism should theoretically not be possible until later in development, when long-term memory and the needed perceptual skills become available. More important, for feelings such as guilt or anger to have evolved for monitoring reciprocal altruism, the circumstances that elicit such feelings must have been prevalent in ancestral environments. Furthermore, a learning mechanism must exist whereby new events, encountered within an individual's lifetime, can come to evoke these feelings. That is, we should not expect a child to feel guilty over failing to repay a sum of money until money itself has acquired the ability to elicit a positive feeling in the child. But when agents or events have acquired the ability to elicit a positive hedonic tone, then we should expect the same range of feelings to accompany gains, losses, or failures to repay this enlarged repertoire of services or commodities. Some of our feelings, then, may regulate reciprocal altruism. But there are still other ways to increase our biological fitness.

Like it or not, we don't live forever! Our bodies and minds may appear to have been designed by natural selection to ensure our personal survival, but this is simply not the case. Certainly we have a sophisticated immune system, and we possess a multitude of affects for regulating such factors as body temperature, blood sugar level, and salt/water ratio. One factor, however, doesn't make sense from a personal survival viewpoint: the sex drive. Based on a study of eunuchs, a human male could extend his life span by about five years through the simple act of castration! For a female, avoiding the perils of childbirth would, on average, also increase longevity, even in our world of modern medicine. The sex drive has, and probably will always have, some detrimental effects on our personal survival. Sexual reproduction, however, is absolutely essential for the survival of our genes. If personal survival were the "goal" of evolution, then sex would feel bad. This, of course, is not the case. Natural selection always favors attributes that enhance the survival of our genes, not simply our own personal survival. We die, but our genes, like diamonds, may go on forever.

Because of the way we reproduce, our genes have a higher probability of being present in close relatives than in distant relatives or strangers. An act of altruism directed toward a brother or sister would therefore incur significantly less biological cost than the same act directed toward a total stranger. This *kin altruism* or nepotism can be viewed as a special case of reciprocal altruism, with costs reduced in accordance with genetic closeness. All things being equal, we should feel more altruistic toward close relatives than strangers and be less concerned when such transactions go awry. The same is true for others. Most of us have benefited from the assistance of kin, and such acts of altruism have certainly made an important contribution to the reproductive success of many individuals throughout the course of human evolution. While living for millions of years in small social groups surrounded by their kith and kin, humans have had ample opportunity to evolve and refine the feelings required for the effective practice of kin and reciprocal altruism. Indeed, given the favorable benefit-to-cost ratio that is inherent in kin altruism, it likely served as the stepping-stone necessary for acquiring and refining the feelings needed for effective use of reciprocal altruism.

Personal fitness, defined as relative reproductive success, is a product of survival to reproductive age, reproduction, and care of offspring, plus the additional contributions that can result from reciprocal and kin altruism. Our repertoire of emotions monitors one or more of these fitness-enhancing domains. With this in mind we can now define emotions as qualitatively different conscious states that have evolved to represent the nature, magnitude, and direction of expected threats or benefits to some aspect of our personal fitness. In many respects, then, emotions are very similar to affects. The difference lies in the fact that emotions are evoked by specific production rules, rather than by simple internal or external sensory events, and their qualities reflect the nature of the potential threats or benefits to our reproductive success.

Like bodily affects, the positive or negative hedonic tone of an emotion is proposed to be a reliable "omen" that foreshadows reproductive consequences and provides the necessary value system for learning to adapt to rapidly changing aspects of the environment. As with affects, the intensity of an emotion magnifies the reproductive consequences of the events that evoke it. These intensity differences can be so dramatic that we often describe emotions by different names when they are really just different degrees of the same emotion. Happiness, for example, runs from contentment to joy to ecstasy, while degrees of sadness may be described as discontent, unhappiness, grief, or even depression. We may report intense disgust as loathing, revulsion, or contempt, and we may describe fear as ranging from apprehension or anxiety to intense panic or terror. Amazement and excitement depict degrees of surprise, and anger may vary from mild irritation to full-blown rage.

These felt intensities, however, are all amplifications of the reproductive consequences of the events that evoke them. A person may feel very sad or happy over the loss or gain of a hundred dollars, even though the actual reproductive consequences may be, on average, a gain or loss of only 0.00001 children! Clearly the felt intensity is out of proportion to its real biological consequences. Recall that this is the same relationship that we discussed with regard to the intensity of an affect, like the pain elicited by a pinprick. Here the intensity of the pain exaggerates the slim possibility of subsequent infection and death. The intensity of affects and emotions can then be viewed as amplifications of the ultimate reproductive consequences arising from the current physical or social circumstances.

To support this viewpoint, it is necessary to show that the environmental and social circumstances that evoke emotions, at least the initial eliciting events, could be expected to reliably and repeatedly occur in ancestral environments. Adaptations can evolve by natural selection only if the events that elicit them remain fairly constant from generation to generation. Second, it is necessary to provide a mechanism whereby new events or agents come to evoke emotions while the emotions themselves remain constant and retain their relationship to reproductive success. That is, since it is proposed that each qualitatively different emotion is evoked by fixed contingencies that reflect the relationship between environmental events and their reproductive consequences, these production rules should remain constant as the specific events that elicit an emotion increase in number. Does a fear of a drop in the stock market involve the same production rule as a fear of strangers? Finally, it is necessary to clarify the role of emotions in learning and reasoning so that such mechanisms are, on average, adaptive. Support for the first proposition can be gleaned from considering the initial contingencies that mediate different emotions in a human child and from examining how each of these unique circumstances is related to the child's biological survival.

THE PRIMARY SOCIAL EMOTIONS

Humans are social animals, and the ability to form social bonds is crucial for their personal survival. Compared with other animals, including other primates, a human child is extremely immature and cannot survive at all without the benefits of a caregiver. From birth a child must enlist and maintain the assistance of an adult in order to meet its most basic survival needs. Without such help the child will perish. Fortunately a child's biological parents also have a stake in their infants' survival since children are the conduits through which parental genes flow into future generations. These cold biological facts are clothed in the warmth of tender, powerful, and persistent emotional feelings that ensure the development of a rapid and stable attachment between a caregiver, usually a mother, and a child.

When it is born, a human child has no understanding of the complex physical and nutritional needs that are required for its survival. It does, however, possess a number of primary emotions (happiness, sadness, anger, fear, disgust, and surprise) that are initially evoked by a small set of commonly occurring events that were consistently present during the long ancestral history of its species. The emotion of happiness, for example, is expressed during the first few weeks of life and appears to be initially elicited by events that satisfy basic bodily needs or indicate the close proximity of a caregiver. Smiles, as a reliable index of happiness, occur when a baby is full, asleep, stroked, or rocked, or in response to a soft high-pitched voice. Irritation of a baby's skin invariably elicits crying and evident displeasure, but gentle touch and rocking movements evoke spontaneous smiles and apparent pleasure. Later, between six and ten weeks, a broad social smile becomes evident, usually in response to a human face. These emotional reactions and their developmental pattern are stable across cultures and are consistent with a child's description of happiness as being evoked by "getting what I want." Other primates, such as rhesus monkeys, also appear to be comforted by soft bodily contact and show a preference for a soft cloth "mother" over a wire "mother," even when the latter has been consistently associated with feeding. In an ancestral environment such soft touch could have been provided only by the closeness of a caregiver, but soft toys and blankets can act as substitutes, and bonding to such objects is not uncommon in the modern world. It appears that happiness is initially evoked by events that satisfy basic survival needs, and it is one of the primary emotions responsible for establishing the first social bond between child and caregiver.

The rapid formation of a strong social bond is of critical importance, not only for the physical survival of a child, but for its future social development. Studies of children in institutions have shown that even when the physical needs of a child are met, the failure to develop an early emotional bond with a single caretaker leads to slow development, withdrawal, depression, and a variety of later developing social problems, such as an excessive desire for adult attention and difficulty establishing social and affectionate bonds with anyone. One forty-year study examined the fate of more than two hundred children who were "at risk" as a result of a disturbed social environment during the first two years of life. Two-thirds of these children showed serious behavioral or learning problems before age ten or exhibited a variety of mental health problems, delinquent behaviors, or unwanted pregnancies by age eighteen. An examination of the one-third who managed to live fairly normal lives revealed that a high degree of socialness and early bonding to a single individual, such as a grandparent or a school-teacher, were the two major protective factors. Given the extreme importance of the child-parent bond, it is not surprising that more than one emotion is involved in forming and maintaining this bond. The two primary emotions of happiness and sadness are expressed early in development and are initially elicited by events that enhance or threaten the child's survival, and they both appear to play a role in the formation of social bonds.

If a child feels happiness when it "gets what it wants"; it feels sadness when its needs go unmet. Children cry frequently during the early months of life, and this apparent unhappiness can be reduced or eliminated by holding, rocking, feeding, or changing a diaper. During this time withdrawal of food or bodily contact can precipitate an outburst of crying. After six to nine months of age, when most children have developed a clear attachment to a specific caregiver, removal of the caregiver will evoke extreme unhappiness, or depression. Even in adults sadness is most intense following the loss of a parent, spouse, child, or close friend. When faced with such disturbing social circumstances, it may be adaptive for adults to temporarily withdraw from social interactions. For a child, however, sadness that does not lead to the attention of a caregiver can be extremely harmful.

For a child, prolonged absence of the caregiver, perhaps as a result of hospitalization, may evoke a deep sadness that has lasting and detrimental consequences. At first the child whines and clings to anyone who offers care, followed by a period of excessive crying and weight loss. Finally, after about three months of separation, the child seldom whimpers but now appears to be withdrawn, depressed, and apathetic. A child appears to possess specific emotions that bias it toward attachment to a single caretaker, and any disruption during the formation of this first social bond can lead to excessive sadness and future avoidance of all such emotional attachments. With persistent and reliable care, such children may show some recovery, but more often than not they enter into superficial relationships and avoid close personal relationships.

The six primary social emotions are all apparent during the first two years of life. These emotions can all be recognized by the facial expressions that characteristically accompany them. Smiles, frowns, stares, and grimaces are reliable indicators of happiness, sadness, anger, and fear. A protruding tongue with closed eyes reveals disgust, whereas an open mouth and wide eyes are reliable signs of surprise. The observation that congenitally blind and deaf children exhibit this same range of facial expressions is a strong indication that these expressions and their associated feelings are, like the basic bodily affects, part of our biological nature.

Children, when old enough to communicate, describe anger as a consequence of physical pain, and fear as arising from a real or imagined physical threat, such as falling or "animal-like" monsters. These two related emotions have much in common and are closely associated with attack (anger) or retreat (fear) in the face of real or anticipated pain. The defensive feelings of fear and anger may have first evolved in response to predators or other dangers that were consistently present throughout our evolutionary past. As noted earlier, many aspects of our environment remain constant over generations, providing an opportunity for the evolution of adaptive solutions. Visual systems, vestibular systems needed to maintain our balance, and circadian rhythms are adaptations that prepare us for life on a planet with an overhead sun, a specific gravitational field, and a constant day-night cycle, respectively. These adaptations are specific to the earth, and we should not expect our hypothetical aliens to possess the same attributes. On the earth microscopic creatures have to struggle with air currents when they fall, but for larger animals it is gravity that poses the major problem. Falling has, and continues to be, a danger to the survival of all large animals that live above the surface of our planet.

It is not surprising, then, that a specific fear of falling can be demonstrated early in a human child's life by using an experimental apparatus known as a visual cliff. This is created by placing a thick sheet of glass on top of a platform, with the glass extending beyond the platform's edge. A child of six months or older, when placed on the glass and encouraged to crawl, will stop at the apparent cliff and display all the behavioral and facial expressions indicative of fear. These specific fears appear to have initially been elicited by circumstances that have posed real dangers to biological survival throughout our long evolutionary history.

Given our prolonged history as social animals, it is not surprising that some social circumstances also evoke a specific fear response. A mild fear, known as stranger anxiety, is evident in most children shortly after they have bonded to their caregiver. At this stage of development, the infant displays fear when suddenly confronted with an unknown adult. In this anxiety-provoking situation infants avoid strangers and cling to their caretaker. Loud noises, like those produced by ancient predators, will also evoke a specific fear in young children. This repertoire of specific fears, however, is only a beginning; many new fears are acquired by a procedure known as classical conditioning.

Conditioning a fear response, using a procedure first described by the Russian physiologist Ivan Pavlov, is a good example of how new environmental events can come to elicit fear (and other emotions). Conditioning a fear depends upon the observation that children between eight and twelve months are easily frightened by a sudden and unexpected loud noise. If this loud noise is repeatedly presented whenever a child is shown a furry animal, the sight of the animal will soon evoke a fear response. This apparently simple associative paradigm can be demonstrated across many different species, including humans, and it is effective in very young children, even though the nervous system is very immature. A closer look at classical conditioning offers several important insights into how human feelings can become elaborated by experience.

When a conditioned fear response is established, presentation of the furry animal does not elicit any conscious experience of the loud noise; it does, however, produce a conscious experience of fear. It may appear that the fear, originally elicited by the loud noise, is now being evoked by the fur, but this interpretation is incorrect. Using Pavlov's paradigm, consider what happens if a painful stimulus replaces the loud noise. Pain evokes a completely different negative feeling from fear. But after repeated associations of fur and pain, the fur by itself will evoke a feeling of fear and not a feeling of pain. The common element between a loud noise and pain is their shared unpleasantness, their negative hedonic tone. It is the acquired association between the furry animal and this negative hedonic tone that evokes a fear response. Fear can be aroused by large variety of events that are predictive of negative hedonic consequences. Adults may fear a stock market crash, the loss of their home, or incurable disease. The common production rule that evokes both specific and learned fears is the expectation of events that have, in our ancestral or recent past, reliably elicited a negative hedonic tone.

The major consequence of classical conditioning is that new events can come to elicit the hedonic tone, but not the specific feelings, of the events with which they have been associated. After conditioning, the negative hedonic tone and intensity of a conditioned fear come to reflect the unpleasantness and intensity of the anticipated pain. That is, the magnitude and direction of the hedonic shift elicited by the furry animal is a function of the magnitude and direction of the feeling originally elicited by the painful or fearful stimulus with which it was associated. This learning paradigm ensures that the intensity of a learned fear is quantitatively related to the real biological threat reflected in the negative hedonic tone of the original event. As a result, we experience more fear of events that signal a large decrease in hedonic tone, than those that signal a small decrease in hedonic tone. Since the events that initially evoke a negative hedonic tone are those that were a real threat to biological survival in ancestral environments, the relationship between feelings and gene survival is maintained as the repertoire of learned fears increases. Under normal circumstances learned fears, just like innate fears, will continue to be reliable omens of potential decreases in reproductive success.

When a new event has acquired the ability to evoke a positive or negative feeling, this relationship may persist for a lifetime. In the case of a conditioned fear developed during childhood in response to the sight of a furry animal, the sight of fur may continue to evoke fear into adulthood. The adult, however, may have no specific memory of the events that led to these disturbing emotions. Psychiatrists spend a great deal of time treating anxiety disorders that may have arisen from chance associations that were formed in early childhood and are now long forgotten. Associations that are formed in a young child, before sensory discrimination is fully developed, will inevitably have a great deal of generalization. Both a child and an adult may not only fear a specific furry animal, but have a generalized fear of all furry objects or even the environmental context in which the fear was first experienced. In the modern world this process appears to be maladaptive, but it had potentially adaptive consequences in ancestral environments. A child bitten by a rat may have been well served to be frightened by all rats and perhaps other similar animals or environments. It may also have been adaptive to maintain this fearful attitude into adulthood.

Pavlov's paradigm is a behavioral description of a neural mechanism whereby events that had no initial value can come to elicit a positive or negative hedonic tone. Such a mechanism permits feelings, initially evoked by a small set of environmental circumstances that have remained stable across many generations, to be aroused by a much larger array of events that vary within individual lifetimes and differ significantly among individuals. Nevertheless, these learned values are not arbitrary, since they were acquired as a consequence of faithful associations with the primary value system. If a single caregiver is associated with the events that initially evoke pleasure, such as gentle touch, rocking, or feeding, then many aspects of this person may soon elicit a similar hedonic tone. The child may now feel happy at the sight or voice of its caregiver. Traditionally psychologists have placed much emphasis on the observation that a behavior (such as salivation) elicited by an events (like feeding) can be transferred to a second stimulus (such as a bell) that has been reliably associated with the event (feeding). But as discussed earlier, nutrients elicit a positive affect, and the transfer of this hedonic tone appears to be the most important outcome of the conditioning procedure.

As the number of events that elicit happiness increases, so does the number that elicit sadness, evoked by the loss or withdrawal of these events. As with all learned feelings, the intensity of this sadness will depend only on the magnitude of the hedonic shift and not on the specific nature of the lost event. If money acquires the ability to elicit a positive hedonic tone, then the loss of money will elicit a comparable degree of sadness, with its negative hedonic tone. As the circumstances that elicit a negative hedonic tone increase in number, so also do the conditions that elicit fear. When confronted with circumstances that indicate a potential financial loss, an individual would experience anxiety or fear, depending upon the magnitude of the expected loss. Our feelings, and the production rules that elicit them, remain constant as the events that satisfy these contingencies increase throughout our lives.

Although classical conditioning has wide applicability, some constraints make it even more adaptive. For example, it is much easier for a rat to associate a sound with a painful stimulus applied to the surface of its skin than to associate a chemical taste with the same stimulus. In contrast, a chemical taste is more easily associated with the internal pain arising from an illness. Pavlov's paradigm appears to be biased toward making associations that are, or were, most adaptive in natural environments. It is much more likely that an inner pain has resulted from something ingested and that a surface pain has been inflicted by an external agent. Such adaptive predispositions to associate are also present in humans. Extreme fears or phobias of spiders, snakes, strangers, open spaces, or heights are much more common than phobias of guns, knives, cars, or other devices that pose a real threat to survival in the modern world. We seem to be predisposed to form rapid conditioned fear responses to events that were a real threat to survival in our ancestral world.

Children first describe anger as a response to physical pain. Indeed, pain aggression is a very common adaptive response that can be observed in many different species. Intense stimulation of the body surface with bites, slaps, or shock will elicit aggression, whereas pains of other origin, such as stomachaches, will commonly evoke fear rather than anger. Very young animals seem predisposed to make a crude distinction between pains arising from the actions of others, pains arising from skin stimulation, and pains of inner origin. A negative hedonic tone arising from the actions of others appears to define the essential production rule that evokes anger. An adult may feel angry, for example, when someone steals their money. For anger to be evoked under these circumstances, the loss of money must elicit a negative hedonic tone, and this consequence must be attributed to the behavior of some other individual. Both of these criteria depend upon a substantial amount of learning. But despite this increase in complexity, the contingencies that mediate anger remain unchanged, the qualitative feeling is the same, and the degree of anger is still related to the magnitude of the loss. Anger, like all other negative feelings, is elicited by an initial set of well-defined environmental circumstances that would have been frequently encountered and that would have posed a threat to biological survival in ancestral environments. Individuals initially feel anger when exposed to these ancestral circumstances, but as new events acquire positive or negative hedonic tone as a function of an individual's learning, the repertoire of evoking events increases as well. As environmental events acquire the ability to elicit positive or negative feelings, they also acquire the ability to support a second form of learning.

BEHAVIORAL LEARNING

The hedonic dimension of feelings can be envisaged as a scale rang-ing from an extremely pleasant or positive pole, to an extremely unpleasant or negative pole. All feelings, in addition to their unique qualitative natures—like disgust, pain, pride, happiness, or sadness—are accompanied by shifts along this common hedonic dimension. Such changes can be viewed as positive or negative rel-ative to the current hedonic state, while the magnitude of a shift determines the intensity of an experienced feeling. Happiness involves a pleasant hedonic tone by virtue of a hedonic shift toward the positive pole of the dimension, whereas the unpleasant he-donic tone of sadness can be viewed as a movement in the oppo-site direction. A child will experience a positive hedonic tone when an event, agent, or circumstance that has elicited a negative emo-tion, like sadness, is terminated. That is, the degree of unpleasant-ness changes from some point near the negative pole to a point that is somewhat less negative. Similarly, the degree of an experienced sadness—from discontent or unhappiness to deep grief or even depression—will vary with the magnitude of the negative shift. For an adult, the loss of ten dollars may elicit a mild unhappiness, whereas the loss of a child may provoke a profound depression.

This shared hedonic dimension is the remarkable attribute that allows our many different feelings to interact with each other. The unpleasantness of fear, anger, guilt, or pain may be greatly reduced by the positive feelings evoked by a parent or a close friend. More important, these positive and negative shifts in he-donic tone define rewards and deterrents, the critical aspect of feelings that enables them to play a collective role in learning mechanisms. Shifts in the positive direction, evoked either by events that elicit positive feelings or by the removal of events that elicit negative feelings, correspond to what psychologists refer to as positive or negative reinforcers, respectively. (Psychologists define reinforcement as events that increase the probability of behaviors that they follow. A reward is a positive reinforcement; *removal* of a deterrent is a negative reinforcement.) Psychologists have known for a long time that a behavior will be learned when it is followed by the presentation of a reward (positive reinforcement) or the removal of a deterrent (negative reinforcement). This learned behavior is stored and will be more likely to occur again at a future time under similar conditions. Alternatively, a behavior will be less likely to recur if it is followed by a deterrent or by the loss of an event that elicits a positive hedonic tone. These simple learning principles have been shown to be effective across many different species. Positive shifts in hedonic tone appear to facilitate behaviors, while negative shifts appear to inhibit them. Hedonic shifts, evoked by events that elicit happiness and sadness, define the mutual reinforcements responsible for the formation of the strong social bond between mother and child. Every smile, frown, or cry reinforces or inhibits the actions of an attentive mother, whose words and touches leave their permanent imprint on the developing child.

These learning principles were designed into the Sniffer simulation. For Sniffers, positive shifts in hedonic tone facilitated behaviors, while negative shifts inhibited them. In this learning paradigm, however, behaviors should not be construed as simple sets of specific muscle movements; they are organized behavioral acts. For example, an animal that has been consistently rewarded for pressing a lever with one hand may suddenly press the lever with its other hand—an act that involves a completely different set of muscle movements. The same behavioral act can be achieved by using many different muscle movements. Indeed, every occurrence of a behavioral act is inevitably different, since it is a coordinated controlled sequence of muscle movements that must be executed in spite of many different body positions and gravitational influences. Gary Cziko has convincingly argued that the only way to achieve constant behavioral acts from a multitude of different starting positions and various intervening disturbances is to use a hierarchical feedback system, like cruise control in an automobile.[2] In this case the driver merely sets the desired speed, and a variety of feedback-controlled subsystems achieve this goal, irrespective of the current speed or obstacles such as hills or tailwinds.

Behavioral acts appear to be organized by the extrapyramidal motor system of the brain. Like setting a cruise control, electrical stimulation within this motor system can repeatedly evoke consistent and organized sequences of behaviors, irrespective of the animal's starting position or interfering obstacles that require new motor movements in order to achieve the same goal. For example, a brief electrical stimulation within the extrapyramidal motor system may cause a monkey to stand up, circle to the right, climb a pole, descend, and attack a subordinate monkey. With regular stimulation the monkey may repeat this sequence of acts thousands of times. This response to brain stimulation is not just a set of fixed muscle movements. In order to complete the entire set of acts, the animal will avoid obstacles placed in its path and search for a subordinate monkey. These directed-response sequences, through which organisms interact with their environment, appear to be the behavioral elements of the motor learning paradigm. Unsurprisingly, therefore, the pleasure pathway of the brain, the MFB, activates the major link between the emotional system and the extrapyramidal motor system: the nucleus accumbens. This junction permits the highest level of organization of behavioral acts, the cruise control settings, to be facilitated or inhibited by positive or negative shifts in hedonic tone. The direction of the hedonic shift dictates the facilitation or inhibition of motor acts, while its magnitude regulates the degree of modification that takes place.

This learning mechanism is effective from the very earliest stages of a child's life. Like natural selection it provides a method for acquiring novel adaptations using selection rather than instruction. In any environment a variety of behaviors may be generated, but as with Sniffers only successful acts will be retained. Success, in this case, is defined by the positive or negative hedonic consequences, rather than by the life-or-death feedback of natural selection. These positive or negative hedonic outcomes become a learned consequence of the stored behavior pattern. When a learned association has been established, then the hedonic outcome will be expected when this behavior is performed at a future date.

BUILDING A SEMANTIC MEMORY

Behaviors are always learned in both external and internal contexts. That is, when an animal learns a behavior in one environmental context, it is likely to exhibit that behavior when it finds itself in the same or a similar environment. This is possible because at the time of learning, a neural representation of the external environmental is present in the brain, and these contextual cues are stored as an integral part of every newly learned association. As a consequence, stored acts are more likely to recur in future environments that possess cues similar to those present at the time of learning. The qualitative aspect of a feeling state, like a qualitative perception of the external world, supplies the important internal context that is also present at the time of learning. Like external context, this internal context is also stored as an integral part of every newly learned association. As a result a behavior learned in one feeling state, such as hunger or fear, is more likely to be recalled in a future but similar internal state. A typical child's memory would then be in the form: "In this environmental situation (external context), when feeling hungry (internal context), drinking a glass of milk (behavioral act) felt good (hedonic consequence)." This is a genuinely *semantic,* or meaningful, *memory,* because events in the physical environment are no longer arbitrary incidents but are connected to the child's welfare through both the internal context and the expected hedonic outcome (Figure 5.1). Without such connections the world simply consists of physical and chemical events that take place inside or outside the brains of living creatures. This physical environment is filled with lawful and consistent events, but such happenings are devoid of all meaning, for the laws of physics and chemistry that govern the behavior of these events have no inherent purpose or intention. The building blocks of meaning can arise only from learned associations between environmental events and the evolved emergent feeling states of conscious biological organisms. A child's developing semantic memory depends upon learning the relationship between those feelings, its behavior, and the world around it.[3]

By acting on its environment, a child soon acquires a repertoire of successful behaviors, the external and internal contexts in which such stored adaptations have been successful, and a set of hedonic consequences associated with the various behavioral outcomes. A child's learning is rapid, and the second year of life is a period of unmistakable cognitive development. Until this time the child's social world has been centered on its mother and her attentiveness. Mother's smiles, laughs, and touch elicit happiness or joy,

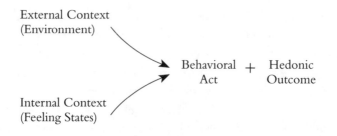

Figure 5.1: The building blocks of semantic memory

while her frowns, inattention, or absence can evoke sorrow or fear. Now the child discovers that it is also an active participant in this relationship and that its actions can influence its own feelings as well as the feelings of its mother. As a consequence of this discrimination between self and others, new secondary emotions emerge.

THE SECONDARY SOCIAL EMOTIONS

The secondary or self-conscious social feelings—guilt, pride, and envy—develop somewhat later than the primary emotions, but they are evident in most children before the age of three. Whereas primary emotions may be common to many species, these secondary feelings appear to be a consequence of the unique history of human beings as complex social animals living in small, tightly knit social groups. For thousands of generations human children grew up within small groups of hunters and gatherers surrounded by their caregiver, many close kin, and many would-be friends. The secondary social emotions may have had their origins in this unique social context.

As discussed earlier, the practice of reciprocal altruism requires a number of highly specific cognitive abilities. Generalized reciprocity is possible only when children have the ability to recognize each other as individuals and are capable of remembering and quantifying many different kinds of goods and services. Quantification and recognition skills, like facial recognition ability, are prerequisites for the effective trading of many different commodities with a large number of different individuals. Although some examples of reciprocal altruism occur in other species, like the blood sharing of vampire bats, generalized reciprocity and the secondary social emotions that monitor such transactions appear to be unique characteristics of human beings.

Between eighteen and twenty-four months, children may hang their heads with embarrassment or shame; they appear to feel guilty about some of their own behaviors. Shortly thereafter they can be seen to beam with pride when they perform a difficult task. Personal achievements are the first source of pride, while actions that elicit negative feelings in the mother define the initial contingencies that mediate feelings of guilt. The ability to recognize the mother's feelings is greatly facilitated by the abundant shared facial expressions that have a common meaning for both parties. These basic expressions, like the feelings that evoke them, are expressed in the absence of any learning. Children's facial expressions are automatic and are controlled by subcortical centers within the extrapyramidal motor system. Although adults develop cortical control over their facial expressions, using their pyramidal motor system, the mother-child relationship has no need for such concealment. The child's first excursion into "mind reading" takes place within the security and honesty of the maternal bond.

Like all feelings, the specific events that elicit pride and guilt become more elaborate with experience. A child's social network soon grows from the simple world of self and mother to a complex community filled with kin, kith, and occasional strangers. In human ancestral environments close contact during early development was probably a reliable cue for identifying kinship, and it still appears to play a critical role in kin recognition. Children who are in close physical contact with other children during their early years, such as those brought up together in Israeli kibbutzim, develop strong sibling bonds similar to those between genetically related brothers and sisters. A kin group may be established by exposure alone, but the development of a kith group requires reciprocal interactions. Armed with feelings of anger and guilt, a child can engage in reciprocal transactions outside of its family. In this manner a kith group of trusted friends begins to develop over the next few years.

During this time the range of pride gradually expands outward from self, to kin, and eventually to kith. A child soon experiences pride in the achievements of its kin and eventually the success of its close friends. The production rule that elicits pride is now much clearer. Pride appears to be first evoked by personal achievements but is eventually elicited by the success of anyone with whom we share genes, or who has helped us in the past. Their success is our success, their happiness is our happiness, and in a biological sense it really is. When our personal actions evoke unhappiness in members of our kith or kin groups, we experience guilt. The more pleasure we have received from an individual, the more pride we seem to feel in their accomplishments and the more guilt when we cause their unhappiness. The negative feeling of envy is reserved for the success of distant acquaintances or total strangers. The unique and biologically related production rules that evoke feelings like pride, guilt, and envy are strong testimony to their biological design.

Social emotions, then, occur early in childhood and are initially evoked by a small set of important events that our ancestors repeatedly encountered and that consistently enhanced or threatened their biological survival during the course of our long evolutionary history. Individuals who could detect and respond appropriately to these important events would certainly have enjoyed more reproductive success than those who were oblivious to such circumstances, favorable or threatening. But for complex animals that live in rapidly changing environments, the ability to expand this value system is of paramount importance. Survival under these conditions depends on the ability to detect and respond to new opportunities or threats that are unique within individual lifetimes. Learning mechanisms are required.

Over a lifetime each individual learns an elaborate new repertoire of events that are capable of eliciting different feelings, but the feelings themselves remain unchanged. These feelings are the ones we were born to experience—the results of millions of years of evolution. A piece of lemon placed in a child's mouth will elicit the facial expression that we easily recognize as disgust. Its eyes close, the corners of its mouth are pulled back, and its head may shake from side to side as it expels the distasteful mouthful by protruding its tongue. For an adult, this same hedonic tone and facial expression, including tongue protrusion, may be elicited by the "disgusting" behavior of a political candidate. The specific events that elicit disgust have undergone a remarkable change, but the similar hedonic tone and expressive features attest to their early origin. Human feelings appear initially to be evoked by a small set of specific events, but as complex animals living in rapidly changing environments, it is the ability to learn and expand this value system that is critical for our survival.

For learning to evolve, however, it must be adaptive. Individuals who learn must, on average, enjoy more reproductive success than those who do not. They achieve this goal by storing environmental events or behavioral acts that are predictive of biologically important consequences. This is the structure of the various learning paradigms already discussed. Classical conditioning transfers hedonic tone from an important biological event to a predictive event, and, by doing so, generates a new set of secondary rewards and deterrents that are quantitatively related to the primary set. Such new rewards and deterrents then provide the hedonic tone and arousal required for storing behavioral acts that can acquire pleasure or avoid displeasure—the omens of reproductive success.

PROBABILITY AND LEARNING

One feeling, however, differs from all others: the emotion that we call surprise. This "colorless" feeling possesses no inherent hedonic tone, yet it plays an important role in enhancing the efficiency of learning mechanisms. Surprise is evoked by unexpected events, and it varies in intensity as a function of the difference between what was expected and what actually occurs. It generates an increase in arousal that varies with the subjective probability of an event. An unexpected gift may evoke more happiness than an anticipated gift, and an unexpected death may elicit more sorrow than an expected death. In both cases, however, the apparent hedonic tone is merely an increase in the intensity of the existing feeling. "Happy surprise" is happiness whose arousal is increased by virtue of its unexpectedness. Surprise has no inherent hedonic tone, but like other feelings it does have its own subjective quality and intensity. Novelty and unexpectedness merely arouse the nervous system, but this arousal is the necessary and sufficient condition for learning to occur.

For a young child, everything is new. Constant novelty plays a major role in building the network of associations that make up its first model of the world. Once such a model is built, however, the formerly novel events are now expected and no longer arousing. Novelty wears off. This loss of arousal as a result of repeated exposure is known as habituation. Once we are habituated to them, events lose their ability to arouse the nervous system and become ineffective in associative learning paradigms. In contrast, events that elicit feelings continue to generate arousal and support creative learning throughout a lifetime. As we learn new associations, surprise becomes most important when our learned expectations are not fulfilled. In an inconsistent environment, old associations must be rapidly modified. Under these conditions surprise, the effective enhancer of arousal, permits the rapid inhibition of old expectations and the rapid acquisition of new ones.

Arousal, resulting from novelty, affects, or emotions, appears to be the common factor that is required for all associative learning paradigms, but the quality of each feeling is also important. Each qualitatively different feeling is evoked in response to a unique set of environmental or social contingencies that have a potential impact on reproductive success. In this respect feelings are just like perceptions; they are qualitatively different subjective codes that distinguish between the relevant and the irrelevant. By associating learned behavioral patterns with these qualitative codes, both internal and external contexts parse our memories. If we learn behaviors to cope with or escape from fear in a specific environment, then we will recall such memories in similar places or under similar fearful conditions. Finally, with the exception of novel events that possess no hedonic tone, the feelings elicited by rewards and deterrents supply the hedonic outcomes of all learned behavior patterns (see Figure 5.1). All learned behaviors have an expected hedonic consequence—they feel good or bad—by virtue of the learning mechanism.

LAMARCKIAN LEARNING?

Imitation and instruction appear to offer alternative means by which individuals could acquire new behaviors from environmental experience. By no known method, however, can the neural pathways activated in a parent's or teacher's brain be directly transferred into the brain of an observer or a student. No outside influence outside can possibly dictate the complex set of neural connections that a student's brain requires in order to generate an original solution to a novel problem. Every form of learning must involve the exploration of hypotheses generated within the brain and the subsequent stabilization or inhibition of those circuits by appropriate rewards or deterrents. In this selection-based learning procedure, the teacher or parent merely provides the appropriate emotional feedback that dictates the success or failure of the student's hypotheses. This feedback need evoke only positive (or negative) feelings, which can take the form of any of a wide range of rewards (or deterrents).

Copying the behavior of another individual (imitation), or listening to a teacher's directions (instruction), can only activate stored knowledge that already exists in the brain of a student. Neither process, by itself, can lead to new learning. But if the activated knowledge is relevant to the current problem, these procedures can decrease the student's search space by dramatically reducing the range of hypotheses that the student must explore and evaluate in order to arrive at a new solution.

In this respect learning is very similar to the acquired secondary immune response. In the immune system a foreign antigen activates preexisting B cells, and these activated cells can then create a rapid solution by generating new variations of their antibodies. By activating a small number of B cells, the antigen decreases the search space of possible antigens and increases the probability of producing a new, more effective antibody. In a similar manner, imitation or instructional processes can activate preexisting memories and by doing so decrease the search space and increase the probability that the student will explore variations of relevant stored information. Such procedures speed up learning, but they may also lead to outcomes that are fundamentally different from the normal creative discovery process. That is, hypotheses generated in a reduced search space are not necessarily variations of the student's stored knowledge with respect to the problem to be solved; they may be hypotheses concerning the verbal or behavioral acts that are most likely to elicit the approval of the instructor. That is, even though the student may generate correct answers under these conditions, the answers may be totally disconnected from all prior knowledge related to the actual task that confronts the student. Imitation and instruction may be fast, but without appropriate safeguards they can misdirect the creative learning process.

In summary, every feeling possesses a specific subjective quality, a hedonic tone, and an intensity. Each of these three aspects of feelings contribute to learning in its own way. The specific subjective quality of a feeling provides the internal context that is stored with all learned behaviors and increases the probability that they will be recalled during similar internal states. The hedonic tone, positive or negative, facilitates or inhibits the behavioral act it follows: a rewards facilitates and a deterrent inhibits motor patterns. The intensity of a feeling modulates the degree of arousal that is required for learning to occur. Learning proceeds slowly when the reward or deterrent is small, but rapid or "flashbulb" memory follows a large change in hedonic tone. As a consequence, we all have vivid memories of where we were when JFK was assassinated or at the time of the *Challenger* disaster. Finally the hedonic consequence of a behavior—a reward or deterrent—is always stored as the expected outcome of a learned behavior. As we will see later, these stored hedonic consequences play an important role in our ability to reason.

Evolution has provided humans with a primary value system of pleasant and unpleasant feelings that are elicited by events that remained stable and were repeatedly encountered generation after generation. Human beings share some feelings, such as pain, with many other species, while others, such as pride, are a consequence of our unique ancestral history as social animals living in small groups surrounded by our kith and kin. These primary values are the core of human nature, but they are inadequate for biological survival in a complex changing environment. Learning is a mechanism by which new events can evoke our inherent repertoire of positive or negative feelings, based on their association with the primary set. This design ensures that the feelings evoked by such new events remain, for the most part, dependable omens of reproductive success. The expanded repertoire of events that elicit positive or negative feelings can serve as the necessary rewards or deterrents for rapidly acquiring an elaborate semantic memory of adaptive behaviors and hedonic consequences. As we will see in the next chapter, this design is built into the very fabric of the human brain.

SIX

The Pathways of Passion

THE NOBEL-LAUREATE ETHOLOGIST NIKOLAAS TIN-
bergen once proposed that four types of questions can be asked
about any behavior: its evolution, its development, its function, and
its mechanism. These questions define four areas of study that complement one another, and we need all four for a comprehensive
understanding of any behavioral adaptation. With respect to human
feelings, Tinbergen's four questions would be: (1) How did feelings
evolve? (2) How do they develop within each individual? (3) What
is their function? (4) How are they implemented within the nervous system? The Sniffer simulation addressed the evolutionary
question. The development and function of feelings were examined in Chapter 5. This chapter is concerned with neural mechanisms.

An examination of the neurophysiology of feelings reveals
that the activation of a major neural pathway, the medial forebrain
bundle, appears to underlie hedonic tone, the shared component
of all feelings. This pathway plays a central role in learning, much
like that proposed and implemented in the Sniffer simulation.
Furthermore, the neural basis of feelings provides a mechanism

whereby emotions can influence both what we think about and how we evaluate those thoughts—our ability to reason.

No single person has had more impact on the neural basis of emotions than Paul MacLean, a senior scientist at the National Institute for Mental Health. His theory, developed over many years, proposed that emotional processing depends on a region of the brain that developed during the evolution of early mammals. MacLean viewed this "old mammalian" region, the limbic system, as sandwiched between a lower, more primitive motor system (the "reptilian brain") and a higher, "new mammalian" system that was responsible for thinking and reasoning. As a consequence the emotional system in primates, like humans, is strategically located between the old motor region of the brain and the recently evolved neocortex of the thinking brain, with reciprocal connections to each of these systems.

The reptilian system, which lies in the central core of the human brain, contains the important extrapyramidal motor system, which is responsible for the many automatic movements, like walking, that are essential for our daily lives. As noted in Chapter 5, a brief electrical stimulation within this motor system will elicit a long sequence of coordinated but stereotyped behaviors. In addition, extrapyramidal damage results in severe motor disorders, like Parkinson's disease. Based on many such studies, MacLean came to view this old "reptilian" region as the control center for unconscious, unfeeling, robotlike motor programs that resemble the behavior of reptiles.

For MacLean, the subsequent evolution of an emotional system in lower mammals was "nature's attempt to provide the reptilian brain with a thinking cap and emancipate it from inappropriate stereotypes of behavior."[1] Over a period of years, he outlined the anatomy of a complete limbic system and divided it into three functional subdivisions, which regulate different kinds of emotional reactions (Figure 6.1). The first subdivision of the limbic system, he said, is responsible for the feelings and behaviors that ensure self-preservation, such as feeding and fighting. The second is concerned with species survival, because it regulates primal sexual functions through feelings and behaviors conducive to mating and copulation. These two subdivisions are interconnected with the old

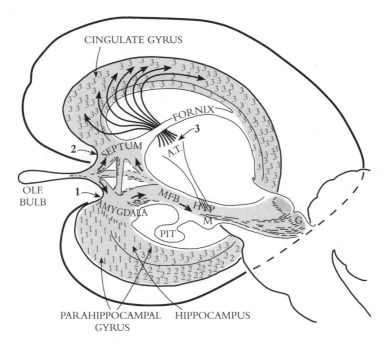

Figure 6.1: The emotional brain: MacLean's "limbic system"[2]

reptilian brain by way of a very large bidirectional pathway known as the medial forebrain bundle (MFB). In addition, each subdivision has a neural output pathway that projects to an area of the primitive emotional cortex, the cingulate gyrus, where such feelings can be experienced. The final subdivision of the limbic system runs from the hypothalamus to yet another region of the cingulate gyrus. According to MacLean, this pathway is concerned with the feelings and behaviors that regulate maternal care and play. In keeping with our earlier discussion of the different factors that contribute to reproductive success, we can view these three subdivisions as being concerned with emotions relevant to personal survival (system one), reproduction (system two), and offspring survival (system three).

MacLean's concept of an emotional brain connected to the thinking brain through its primitive cortex, the cingulate gyrus, and linked to the old motor brain by the MFB, was an attractive idea. First, it provided a mechanism through which the stereotyped behaviors of the old motor brain could be controlled by emotional input from the limbic system. Second, it revealed how the thinking and emotional brains could communicate through the cingulate gyrus, allowing cognitive production rules to elicit emotional reactions and permitting emotions to influence cognitive processes. Third, it provided pathways to the hypothalamus, through which emotional processes could influence the physiology of the body, such as changes in heart rate or blood pressure. Finally, MacLean showed that all sensory inputs ultimately ran to the hippocampus, an integral part of the limbic system, allowing feelings to be evoked by inputs from all of the senses. MacLean's theory, therefore, provided the first comprehensive picture of the brain mechanisms involved in controlling the various aspects of emotional processing. In recent years the theory has been further refined by a series of systematic studies that have focused on the inputs and outputs responsible for a specific emotion: fear.

Research on fear dates back to 1937, when two neurosurgeons, Heinrich Kluver and Paul Bucy, removed the temporal lobes of rhesus monkeys and made a remarkable discovery. Kluver described the result as probably "the most striking behavior changes ever produced by a brain operation in animals." The result (now known as the Kluver/Bucy syndrome) included a complete loss of fear and anger, and a failure to distinguish between appropriate and inappropriate food and sexual stimuli. These animals lacked evaluative discriminations. They would examine the tongue of a hissing snake, nibble at their own feces, continuously manipulate their genitals, and exhibit no vocal or facial signs of anger. Joseph LeDoux, author of *The Emotional Brain,* (1996), saw the findings of Kluver and Bucy as an opportunity to unravel the neural circuitry involved in fear.[3]

Animals, like rats, have a well-defined fear reaction that includes behavioral responses, like freezing, and physiological components, like changes in heart rate. If the sounding of a tone is repeatedly associated with a shock to the animal's foot, then sounding the tone will evoke a fear response that these measures can detect. Systematically destroying different areas or pathways within the animal's brain makes it possible to trace the circuits responsible for evoking and organizing this fear response.

    First LeDoux demonstrated that the critical structure for elic-
iting a fear response is the amygdala—the central part of the survival portion of MacLean's limbic system (Figure 6.1) and the key
brain structure destroyed in the rhesus monkeys that were examined by Kluver and Bucy. Whether a fear is innate, like the rhesus
monkey's fear of snakes, or learned, evoked by a simple tone,
removing the amygdala seems to totally eliminate it. LeDoux then
examined how sensory inputs reached the amygdala and came to
the conclusion that there were two major routes and that each
played a different role in determining the specificity of a learned
fear response.

    Neurophysiologists have known for many years that neural
inputs from sensory organs, such as the ears or eyes, first run to specific thalamic nuclei and then project to specific sensory areas of
the neocortex. LeDoux demonstrated, however, that sensory projections from the thalamus to the cortex were not necessary in
order to evoke a learned fear response to a simple tone. An alternative "fast" pathway, he discovered, ran directly from the thalamus
to the amygdala, and this pathway could elicit fear even after the
thalamo-cortical projections were destroyed (Figure 6.2). Thalmo-
cortical pathways from the senses also reach the amygdala eventually, but they are much slower than the direct route from the
thalamus. LeDoux then examined the function of these slow pathways by teaching animals to fear one tone and ignore another.
Destroying the slow thalamo-cortical route interfered with this
delicate discrimination, causing the animals to exhibit a fear
response to both events. Cortical processing appears to be necessary for selective discriminations, but in its absence it's better to be
safe than sorry!

    Another important insight arising from LeDoux's research
concerns the role of the hippocampus, which also receives input
from the thalamo-cortical pathways activated by our senses. In the
visual system, for example, information from the eye goes first to
the thalamus and then to the visual cortex found in the occipital
lobe at the back of the brain (Figure 6.2). This thalamo-cortical
pathway, however, can be divided into two different streams: one
concerned with "where," and one concerned with "what" is present in the visual input. The "where" system, which processes the

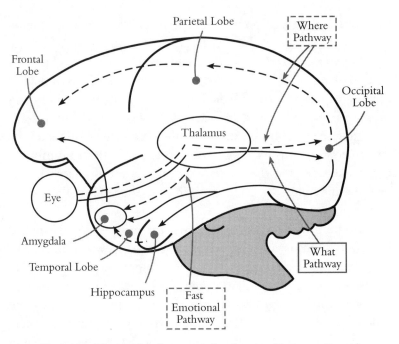

Figure 6.2: The visual thalamo-cortical pathway and its interaction with the emotional pathways of the brain

location and movement of objects, eventually reaches the frontal lobe after passing through the parietal lobe. In contrast, the analysis of shape and color, the "what" information, travels from the visual cortex to the temporal lobe of the brain. Like "where" information, "what" information also eventually reaches the frontal lobe, but it simultaneously dives subcortically into the amygdala and hippocampus. The amygdala input is the slow perceptual route discussed above. But what is the function of the hippocampal input?

LeDoux addressed this question by removing the hippocampus. Animals that have acquired a fear response to a tone often freeze when they are returned to the experimental chamber where the training took place—a fear response to a contextual stimulus. LeDoux demonstrated that this could be eliminated by destroying the hippocampus. Indeed, the hippocampus, which receives cortically processed input from all sensory systems, appears to supply important contextual cues to the amygdala. Lesions of the amygdala eliminate all varieties of fear responses, but hippocampal lesions selectively abolish contextual fear. These and other studies now suggest that MacLean's hippocampal inputs are more involved in the processing of contextual cues than in the processing of emotions per se.

LeDoux's research has established that all fears—innate, learned, or contextual—depend upon activation of the amygdala and as a result share a common set of neural output pathways that originate in the amygdala and transmit (or carry) nerve impulses to a variety of different systems. One output from the amygdala projects to the hypothalamus and is responsible for the physiological reactions to fearful events. Other outputs arouse the neocortex of the thinking/reasoning brain by activating both the nucleus basalis, the important arousal center that is degenerated in Alzheimer's patients, and a brain stem nucleus known as the locus coeruleus. These two pathways allow emotional events to modulate cortical excitability. Finally the amygdala communicates with the old reptilian brain through the MFB, the large fiber tract described by MacLean as the major link between the emotional brain and the old reptilian brain. It is also the pathway that appears to underlie hedonic tone.

The MFB interconnects the emotional and motor brains, with about one million fibers running in each direction. Ever since biopsychologist James Olds (at McGill University in Montreal) made an accidental discovery, the MFB has acquired a totally new significance. Olds discovered that in order to gain electrical stimulation of this bundle of nerve cells, animals would perform a variety of tasks—the stimulation acted just like a reward. Although stimulation of other areas within the limbic system can achieve a similar effect, these other areas actually activate the MFB, which is now generally considered to be the "hottest" zone for self-stimulation. As a consequence the MFB has become known as the "pleasure pathway" of the brain. Its importance lies in the fact that it provides a biological basis for the role of emotions in learning and reasoning. As we have seen, hedonic tone (pleasantness or unpleasantness) is the shared element of all feelings—affects and emotions—and distinguishes them from all other types of subjective experiences, such as thoughts and perceptions. Output from the limbic system, the pleasure pathway, appears to be the common neurophysiological correlate of hedonic tone, and this output establishes the important link between feelings and learning.

A great deal of research has focused on studying the effects of stimulating the pleasure pathway, the major neural output pathway from MacLean's limbic system. This pathway, it is now known, activates a large group of nerve cells that send messages back to the limbic system, and most importantly, they also stimulate the old reptilian brain by releasing a chemical, dopamine, onto the nucleus accumbens. Incredibly, the release of dopamine onto the nucleus accumbens has now been established to underlie almost every form of pleasure that animals can experience. Blocking the action of dopamine at the nucleus accumbens will stop an animal from stimulating its pleasure pathway (in the experiment discussed earlier), and even more important, it will also block the pleasurable effects of a host of natural rewards, such as food and water and sex. In addition, it has now been established that all addictive drugs—like cocaine, amphetamine, alcohol, and heroin—either directly or indirectly activate the pleasure pathway and eventually release dopamine onto the nucleus accumbens. The release of dopamine onto the nucleus accumbens, then, appears to underly all of our rewarding feelings.

For many years psychologists have known that rewards and deterrents play a major role in regulating learning. An animal will learn a behavior that is followed by a reward, like food, or the removal of a deterrent, like a foot shock. These procedures, known as positive and negative reinforcement respectively, have been shown to control learning across an enormous variety of species. Drug addiction in humans demonstrates just how powerful this type of learning can be. In the case of addiction, a drug like cocaine causes a massive release of dopamine onto the nucleus accumbens, and the user experiences a sudden "rush" of extreme pleasure. This euphoria rewards the drug-seeking behavior and leads to the very rapid and strong learning that we call addiction.

Less dramatic, but perhaps more important, is the observation that a wide variety of behavioral acts, when followed by stimulation of the pleasure pathway, will be learned. That is, an animal can be taught to press a lever, run a maze, pull a chain, or do whatever, if performing that act is followed by electrical stimulation of its pleasure pathway. Activation of this pathway is clearly not responsible for learning any one specific behavioral response: rather, it appears to facilitate any number of different behavioral acts. This arrangement was built into the Sniffer model, where positive affect stabilized the synaptic weights within the simulated neural network that governed Sniffers' movements across the computer screen. The pleasure pathway's control of the motor system is also consistent with the earlier theoretical speculation that feelings initially evolved as mechanisms for regulating the automatic approach and avoidance behaviors of less complex organisms. With this pathway in place, the stereotyped behaviors of the old reptilian brain are no longer seen as automatic, and behavior is now considered to be under the control of output from the emotional brain: the limbic system.

A deterrent, like a foot shock, will also release dopamine onto the nucleus accumbens, and a drug that blocks the action of dopamine will also interfere with learning to avoid such a deterrent. The release of dopamine appears to be highly correlated with both positive and negative hedonic tone, although different neurons and receptor sites may be involved in each case, and the effects on the motor system may be either excitatory or inhibitory. In the Sniffer simulation positive hedonic tone (or reduction in negative hedonic tone) increased the future probability of the behavior that led to the reward, whereas negative hedonic tone (or reduction in positive hedonic tone) inhibited prior behavior and generated the exploration of new behaviors. Taken together, these two mechanisms are the basis of the proposed selection-based learning mechanism. Feelings, associated with negative hedonic tone (or the loss of positive hedonic tone), serve as behavioral hypothesis-generators, whereas reinforcements serve as evaluations of these hypotheses by retaining any behaviors that are effective. In this manner a hungry animal will explore a variety of behavioral hypotheses but store only the solutions that lead to food and the associated feelings of reward.

The limbic system may originally have evolved to provide a mechanism whereby survival needs of an organism could regulate and guide the automatic behaviors of the old reptilian brain.

Sandwiched between the motor brain and the rapidly evolving neocortex of the thinking/reasoning brain, however, the limbic system also offers a readily available means for guiding cognitive processes and biasing reasoned decisions in an adaptive direction. In response to potential threats or benefits to biological survival, emotional neural outputs from the limbic system can activate cognitive processes. That is, emotions, just like affects, can cause us to generate many different hypotheses about ways to avoid an expected deterrent or obtain an expected reward. For example, the fear of losing our job may motivate us to explore a wide range of strategies that could avert this negative outcome, such as working longer hours or securing alternative employment. Alternatively, the excitement of a forthcoming date could motivate us to wash the car, dress up, buy a bottle of wine, and make reservations at a restaurant. Outputs from such decision processes can then feed back to the limbic system to be evaluated by the circuits of the emotional brain. In either of the above scenarios, for example, we may generate a large variety of possible strategies, but we implement only those that evoke a positive or reduce a negative feeling. Limbic-neocortical circuits thus act as a closed loop that is capable of continuously generating and evaluating hypotheses in the absence of any sensory input. Such an arrangement permits us to generate long sequences of thoughts that are constantly steered and evaluated by our feelings. Just as with the motor learning mechanism, we can generate innumerable variations of hypotheses until our emotional value system deems a particular outcome to be favorable. We call such activities thinking or decision making, and it is these cognitive operations that dominate our conscious mind.

As we will explore more fully in Chapter 8, our behavioral learning and cognitive reasoning mechanisms have a great deal in common. In both cases we are generating and evaluating a (behavioral or cognitive) scenario, and in both cases it is feelings (affects or emotions) that regulate both of these processes. Like behavioral learning, feelings determine what we think about, and feelings evaluate these potentially new and creative hypotheses. From this perspective behavioral learning and reasoning are both selection-based procedures that depend upon generating innumerable variations of previously stored adaptive solutions, motivated and evaluated by our feelings. Just as with evolution, both processes depend upon conserving past successes and exploring new creative modifications centered on the previously stored solutions.

This brief review of the neurophysiology of emotions has sketched the major anatomical pathways that are activated by emotional events and a mechanism by which these pathways can control organized patterns of behavior and physiological responses. To summarize: The neural processing of emotions begins within the limbic system, a number of interconnected subcortical regions around the hypothalamus. One major neural output pathway from this system, via the hypothalamus, appears to be responsible for physiological adjustments to the body, such as changes in heart rate or blood pressure. A second output pathway, common to all feelings, is the pleasure pathway, which ultimately releases dopamine onto the nucleus accumbens. This pathway is closely associated with the control of motor behavior and is correlated with the hedonic tone that is a fundamental aspect of all feeling states. Limbic outputs also influence two major arousal systems of the brain, and these pathways provide a mechanism by which feelings can modulate cortical arousal. Finally, specific portions of the limbic system project to different areas of the cingulate gyrus, allowing different patterns of limbic activity to generate qualitatively different emotional feelings and bias cognitive processes. We have certainly learned a great deal about the emotional pathways within the brain, and such information has provided insights for understanding both drug addiction and the function of human feelings. Nevertheless, some major questions remain unanswered. How can the actions of nerve cells give rise to our inner private conscious emotions? Indeed, how and why are we conscious of anything at all? These are the most difficult questions of all.

THE NATURE AND ROLE OF CONSCIOUSNESS

No one currently knows what type of neural organization is responsible for our inner conscious experiences. Some intuitive constraints, however, can limit our search to a smaller number of possible mechanisms. If conscious experiences are an emergent property of the nervous system, as discussed in Chapter 1, then we shouldn't expect them to be localized to any one region of the brain. To return to the automobile analogy, seeking the location of a conscious experience would be equivalent to seeking the location of an emergent property like acceleration. Clearly no single part of a car is the sole location of acceleration. Yet although acceleration can't be localized, the potential parts, and the types of interactions between those parts, that are required for acceleration can be identified. For example, an examination of the car might reveal that some parts, such as the pistons, appear to be important for acceleration, whereas other parts, like the seats, seem to play no role at all. Once we have determined the relevant parts, it may then be possible to describe the interactions between them that ultimately give rise to acceleration. Perhaps a similar strategy could uncover the physical basis of conscious experience.

As we have seen, many brain processes occur in parallel. For example, emotional processing, and the "what" and "where" streams of visual processing, involve completely different areas of the brain. We, however, have no conscious awareness of this division of labor. We experience "a beautiful red setting sun," totally unaware of the fact that the pleasantness is processed by different portions of the limbic system: the roundness and redness of the sun are processed by pathways that pass through the temporal lobe, while the position of the sun involves activity in the parietal lobe. This parallel processing creates a major puzzle for any theory of consciousness. How can such spatially distributed operations give rise to a single conscious experience?

Francis Crick, famous with James Watson for discovering the structure of DNA, calls this unexplained faculty of consciousness "the binding problem."[4] The challenge for neuroscientists is to figure out how these spatially separate processes can be bound together to give rise to a unified conscious event. It is tempting to propose that the parallel streams must converge onto a single network, or neuron, that can somehow bind together the information arriving from these many different systems. Indeed, as we have seen, the prefrontal cortex receives input from the "where" and "what" visual systems and is also intimately connected with the cingulate gyrus of the limbic system. Does the solution to the binding problem lie in the prefrontal cortex?

A single cell certainly cannot give rise to a conscious experience. If it could, then we would require either a different nerve cell for every unique conscious experience, or a single neuron that could exhibit many different operating states according to its inputs. Neither solution is possible. Our brain simply does not have enough nerve cells for a different nerve to be associated with every different conscious experience. Furthermore, if each conscious experience depended on the activity of a single nerve, then it could easily disappear since thousands of nerve cells die every day!

Perhaps a more complex circuit of neurons gives rise to conscious experience, solving the binding problem. This circuit could have many different states, and with some redundancy in the network, it could survive the death of many cells. In order to cope with the conjoint problems of multiple representations and ubiquitous cell death, clearly, such a network would have to be quite large. But postulating that a large complex network is responsible for conscious experience brings us back to the problem we started with. We are building a brain within the brain—and our "little brain" has exactly the same problems as the big one! How can the activity of a nerve cell in one area of our new proposed network be bound to the activity of nerves in another area of the same network so that different conscious states can arise according to whether these isolated areas are simultaneously active? The same problem arises when we try to explain how the emotional system knows what is going on in the visual system so that the two phenomena can be bound together to form a single conscious experience. In searching for a way to unify the parallel pathways, then, constructing another "little brain" doesn't get us anywhere.

A wealth of other phenomenological evidence suggests that consciousness does not belong to any single autonomous region of the brain. First of all, conscious experiences are qualitatively different from one another. Seeing, feeling, thinking, and hearing are very different subjective states. Visual consciousness is impaired by lesions of the visual cortex, and hearing is impaired by auditory cortical lesions, so it is likely that consciousness is not an attribute that is exclusively possessed by any one single region but is a property that exists within many areas of the brain. Of course, removing the eyes will also produce a loss of visual consciousness, so we cannot be absolutely sure that visual consciousness is inherent in the visual cortex.

If one particular region of the brain were responsible for the many different qualities of consciousness, it would have to maintain the identity of its inputs in order to generate different types of experiences according to their source. This again would be similar to rebuilding what already exists. It is simply more parsimonious to attribute visual consciousness to visual areas, auditory consciousness to auditory areas, and so on. The experimental evidence also supports this interpretation. For example, Antonio Damasio and his colleagues (at the University of Iowa College of Medicine), have studied patients who experienced a total loss of color vision, achromatopsia, following damage to a very specific area of the visual cortex.[5] What is fascinating about these patients is not the loss of color vision per se, but the fact that they cannot recall or even imagine a color, like redness, that they had commonly experienced before the brain damage. Damage to other brain regions does not produce the same deficit, so it appears that the conscious experience of color depends upon the integrity of a specific region of the visual cortex. Presumably auditory consciousness depends on particular areas of the auditory cortex, and feelings depend on specific regions of the cingulate gyrus. This leads us to the conclusion that consciousness is the result of a dynamic organization that can exist within many different areas of the brain. Certainly other operations, such as turning consciousness on (or off), or changing the contents of consciousness, may involve specific neural areas. That is, the controls for consciousness may depend on highly localized regions of the brain, but the dynamic organization of the brain that underlies all conscious experiences appears to be an attribute that is not imprisoned within any single autonomous brain region.

If you are like me, you have often driven to work with no conscious awareness of how you got there! During the drive your conscious mind may have been preoccupied with planning an upcoming trip or preparing for an important meeting, and you may not be consciously aware of the world around you. Clearly, however, in order to navigate all the turns and signs along your route, your brain processed a great deal of visual information. But it all occurred in the absence of any conscious awareness. An important signal, such as a loud car horn or some another emergency situation, would have automatically provoked a rapid and conscious assessment of your current environment. This suggests that conscious awareness can be evoked by a higher level of arousal, triggered by a high-intensity, unexpected, or important sensory event.

In the nervous system arousal entails an increase in the sensitivity of a large number of widely distributed nerve cells. Under aroused conditions active cells can have a greater effect on their connected neighbors, and hence they can have an expanded sphere of influence. In the presence of increased arousal, it becomes possible for simultaneously active pathways to interact with each other in a manner that would be impossible in the absence of arousal. For example, when we are aroused, the spatially distributed "where" and "what" pathways that are simultaneously active could interact with each other by virtue of their increased spheres of influence. The many autonomous brain regions that process the color, shape, pleasantness, and movement of the sun could likewise interact to produce the unitary conscious experience of a beautiful red sunset. Indeed, the organization of the brain is conducive to exactly this type of interaction.

Brain architecture is characterized by abundant reciprocal connections between cortical regions, recurrent pathways that permit feedback and reactivation of active areas, and lateral inhibition that focuses neural activity within active centers by inhibiting less active adjacent regions. This organization permits neural interactions between active cortical regions to be sustained by reciprocal pathways interconnecting such active regions. Such temporary modes of neural interaction will vary according to the particular brain regions involved, the degree of arousal, and the inherent organization of the reciprocal pathways. A good analogy for such reciprocal activity might be the oscillation of a guitar string. Although the individual particles of the string are unaware of each other's behavior, with every pluck, they rapidly self-organize to produce a characteristic fundamental frequency. The vibration of the string is an emergent property of the whole system. The study of complex systems that involve nonlinear interactions between simple elements has revealed that such systems can indeed give rise to phenomena that cannot be explained by examining the laws obeyed

by the individual components. Such emergent phenomena orga-
nize themselves according to their own microlaws and arise without the benefit of any central controller.

Similarly, cells within a complex neural network may have no
knowledge of the activity of other cells, but each such network may
possess its own preferred mode of oscillation that is a function of
its cellular arrangement. Using the string analogy, it is the resonant—or characteristic—mode of oscillation between excited
brain areas that defines the nature of each conscious experience.
This type of neural organization may underlie our conscious experiences. From this viewpoint, consciousness is an emergent property arising from the self-organization of concurrently active but
spatially distributed regions of the brain; there is no central orga-
nizer and no unique location where it comes into existence. Like
the acceleration of a car, each unique experience depends on the
interactions between many different component parts, and no single part is by itself responsible for the emergent attribute.

This view of consciousness is similar to that proposed by
Susan Greenfield, a professor at Oxford University. In her recent
book *Journeys to the Centers of the Mind,* she describes her "con-
centric theory of consciousness" as follows.[6]

> Consciousness is spatially multiple yet effectively single at
> any one time. It is an emergent property of non-specialized
> and divergent groups of neurons (gestalts) that is continu-
> ously variable with respect to, and always entailing, a stimu-
> lus epicenter. The size of the gestalt, and hence the depth of
> prevailing consciousness, is a product of the interaction
> between the recruiting strength of the epicenter and the
> degree of arousal.

Greenfield views each conscious experience as arising from a
temporary neural gestalt that forms and then gives way to yet
another neural gestalt. Gestalts form around epicenters arising from
a sensory input, an unconscious process, or the output of an earlier
gestalt. From this viewpoint, consciousness is an emergent property that depends upon activity within large neural gestalts, whose
size is regulated by arousal. The formation of a neural gestalt large
enough for a momentary conscious experience, she argues, may be

so large as to preclude the simultaneous formation of a second gestalt. In Greenfield's words, "If in a group of fifteen people, eleven are recruited for a football team, there are not enough to make up a second team simultaneously."

Building on Greenfield's ideas, I would propose that output from the limbic system provides the major source of arousal responsible for the formation of large neural gestalts. Cortical arousal is a common element of all feelings, whether they are generated by sensory inputs (such as pain) or the activation of emotional production rules (as with pride). As a consequence a sensory or social event that stimulates limbic circuits will invariably enhance cortical arousal in the same way that intense or unexpected events do. Therefore any thought, memory, or sensory input that activates the emotional circuitry of the brain will inevitably evoke a transitory increase in cortical arousal, causing such events to enter and dominate our conscious awareness.

The contents of conscious are prioritized by our feelings. During our "autopilot" drive to work, our conscious mind may have been preoccupied with an upcoming presentation, the health of a child, or an anticipated encounter. One thought leads to another, as countless potential future scenarios are generated and evaluated under the constant activation and feedback from our emotional brain. At this point consciousness is directed inward, and we may have no conscious awareness of our external surroundings. This state of consciousness may suddenly be interrupted by the screech of brakes or the sound of a police siren—events that rapidly seize and redirect out conscious attention. The serial flow of consciousness is suddenly interrupted. While this interference may appear to be a departure from the mind's "standard operating procedure," it is not, for whether our conscious mind is processing internal or external events, neural activity is always emanating from the limbic system, which prioritizes these events and dictates the moment-to-moment contents of our conscious mind.

From this perspective, each conscious experience is a consequence of the widespread and almost infinitely numerous potential resonant circuits that can be generated by binding together the many spatially distributed and otherwise isolated parallel processes of the nervous system. Consciousness permits shape, color, movement, depth, feelings, and so on; to be combined into a unified perception such as "a beautiful red setting sun" or "a noxious, brown, heap of manure." The number of possible conscious experiences that can be derived from such combinations of elements is almost infinite, and conscious organisms benefit from their ability to distinguish between each of these different states. Such discrimination is the essential role of consciousness. Of course, no such pictures, sounds, smells, tastes, or feelings really exist in the external world. Each momentary experience is merely a virtual representation that amplifies and discriminates between those aspects of the physical or social world that are biologically relevant. Our conscious mind imposes a structure on our experience of the world around us; it is certainly an illusion, but it is an adaptive illusion.[7]

On the basis of the favorable survival consequences that result from unified conscious experiences, natural selection has evolved a mechanism that binds together many spatially distributed neural processes. The evolution of such a binding mechanism is demonstrated by examining the outcome of the iterative car race discussed in Chapter 1. When engineers repeatedly select racing cars on the basis of a single emergent property like acceleration, they are also selecting for an arrangement and interaction between the parts that contribute to, and augment, this single emergent characteristic. Many of these improvements in performance undoubtedly involve feedback circuits that can adjust the performance of one part—for example, the carburetor—in response to the activity of another part, like the speed of the engine. As a consequence of better coordinating these different parts, acceleration would increase over generations. A comparable outcome would result if engineers selected on the basis of a second emergent property, like cornering ability, as well. In real car races, however, cars that possess both acceleration and cornering ability are the ultimate winners. To accelerate during cornering requires a suspension system that will combine these two different properties into a higher-level system. Such cars would adjust their cornering according to their acceleration, or decelerate whenever they are cornering. By selecting on the basis of overall performance, engineers are inevitably selecting for a higher-level control system that could coordinate the interactions between many different emergent properties so as to optimize that overall performance. Similarly, when natural selection favors the overall behavioral outcomes that result from the coordinated actions of many different emergent properties, it is also selecting for the multitude of unified conscious experiences that can result from this organization.

The purpose of this discussion has been to propose that the dynamic neural organization underlying consciousness has arisen because natural selection has always favored any emergent property of the nervous system that consistently contributed to biological survival. At this point in human history, no one knows the actual physical basis of consciousness but the specific details are not as important as the central proposition—that conscious experiences, like sensations and feelings, evolved because they dictated a dynamic organization of the nervous system that could prioritize experiences and distinguish between environmental events or circumstances that have had a real influence on biological survival.

A television producer once told me that I could make my theory of feelings more understandable to an audience by making a simple statement such as "Orgasms are nature's way of telling you that you are increasing your reproductive success." This sounds simple and clear, but it may also be misleading. If we interpret this statement to mean that natural selection has led to the evolution of feelings, like orgasms, that are evoked by circumstances that have consistently presented opportunities (or threats) for biological survival in ancestral environments, then we would be right, but we'd only be seeing half the picture. What's missing is mention of the functional consequences that arise from this arrangement. Consciousness did not evolve to provide humans with "pretty pictures, sounds, or feelings"! It is not simply the subjective experience of an orgasm that is important, because natural selection can't "see" or act directly on such inner feelings. Whether a person does or doesn't have feelings is irrelevant, unless having feelings changes behavior in a beneficial manner. Perhaps we could say "Orgasms are nature's way of getting you to increase your reproductive success," if we mean the design is functional rather than purposeful. It is the favorable behavioral consequences of feelings that are significant, because these behavioral outcomes specify who survives and who does not.

No doubt David's brain generated a host of pictures and feelings, but his fearful monsters were not favorable emergent properties; they were detrimental to his survival. A detrimental or nonfunctional characteristic, like the "noisiness" of a car, is destined to extinction over generations. Indeed, this has happened with the appearance of electric cars. For an emergent property to persist over generations, it must have immediate and useful behavioral consequences. It is not sufficient to point to the ultimate goal, gene survival, as the single functional outcome that defines which emergent properties survive. Biological survival is certainly the ultimate reason for all such properties, but a comprehensive theory of feelings should explain the immediate useful effects of feelings so that animals with feelings can survive, find a mate, and reproduce. This is the topic of the next chapter.

SEVEN

∞

Mirror, Mirror, on the Wall . . .

WE ARE ALL FAMILIAR WITH THE METAMORPHOSIS OF a tadpole into a frog. This dramatic change in the structure and behavior of the amphibian, whereby a juvenile turns into a sexually active adult, is triggered by the release of thyroid-stimulating hormone releasing factor (TSH-RF) from a maturing hypothalamus. In humans the metamorphosis of puberty is also initiated by a hypothalamic hormone, known as gonadotropin releasing factor (Gn-RF). In both amphibians and humans, these releasing factors are transported from the hypothalamus to the master gland of the body, the pituitary, where they trigger the release of tropic hormones—TSH and gonadotropins, respectively. The tropic hormones then act on other glands—the testes or ovaries, in humans—whose secretions modify the structure of both the body and the brain. The pubertal metamorphosis of a child into a sexually reproducing adult is every bit as dramatic as the transmutation of a tadpole into a frog.

Puberty marks the beginning of the reproductive phase of life, and the biological stakes are higher than ever. Gonadal hormones flood our brain, and the hedonic tone of our feelings

129

increases dramatically, sometimes to orgasmic levels. Of course, since birth, the intensity of feelings has always been regulated by blood factors. Food tasted more pleasant when our blood sugar level was low, the comfort of a warm fire was enhanced by a low blood temperature, and the intensity of pain varied with blood endorphin levels. These blood factors, monitored by structures of the brain centered around the hypothalamus, constantly influenced our sensitivity to environmental factors. Such modest changes in feelings have been a part of our daily existence since the day we were born. None of this experience, however, prepares a child for the dramatic changes that mark his or her entry into the reproductive stage of life. There teenagers experience a metamorphosis of body and mind.

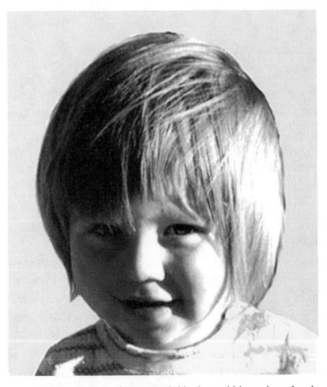

Figure 7.1: Face of a young child who could be male or female

PHYSICAL ADAPTATIONS AT PUBERTY

The physical changes at puberty are obvious and indisputable. Before puberty the faces and bodies of males and females are quite similar, and it is often difficult to distinguish between the sexes (Figure 7.1). Nonreproductive boys and girls enter puberty with almost identical proportions of muscle, fat, and bone. But they exit their metamorphoses as mature reproducing adults with completely different body shapes and compositions. During puberty a young woman gains about thirty-five pounds of body fat—twice as much as a young man. Her mammary glands undergo radical structural modifications, and strategic fat deposits result in pronounced changes in her body contours. At the same time a young man acquires about one and a half times as much muscle and skeletal mass, and both facial and axillary hair are usually evident. Physiologically his metabolic rate, maximum VO_2, and lung capacity increase over the same features in a female. During puberty the male body undergoes a pronounced adolescent growth spurt controlled by androgens, especially testosterone, in the presence of growth hormone. In females a small amount of androgens released from the adrenal glands is presumed to be responsible for their smaller growth spurt. The difference in androgen levels between the sexes appears to underlie the taller height and longer lower jaw measurements found in adult males compared with adult females (Figure 7.2).

To state that men are taller than women does not mean that all men are taller than all women. There are, of course, many individual differences, and our attention is often drawn to these extremes. Yet despite our curiosity about bearded ladies, few of us ask why there is a sex difference in facial hair in the first place. Why is our nose above our mouth? Why are our arms as long as they are? These questions about the ordinary and average are seldom asked, for we consider them to be "the way things are," and we often fail to ask why things are the way they are. For evolutionary psychologists, however, questions about the average and ordinary are of paramount concern. The average often reflects what is, or was, most adaptive for biological survival, and individual variations are

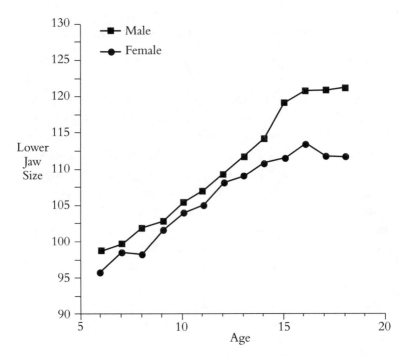

Figure 7.2: Growth of the lower jaw (nasion to chin) in males and females

best regarded as potentially important explorations around this successful theme.

An examination of the features and proportions of average adult male and female faces reveals a large number of differences, even when adjusted for overall body size. On average, men have a more pronounced brow ridge, more sunken eyes, and bushier eyebrows set closer to the eyes (Figure 7.3). Both the nose and the mouth are wider in the male face, while the lower jaw is both wider and longer than that of the average female face. What can account for all these differences given the fact that they appear to have little importance in today's world?

The ten-thousand-year history of the human species in agricultural, industrial, and technological societies is insignificant compared with the thousands of generations during which women foraged for food and men scavenged or hunted for wild prey. Given this long ancestral history, we should expect many attributes of our bodies and minds to reflect the characteristics that were most use-

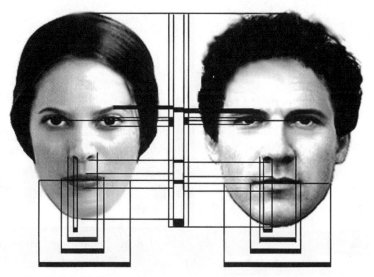

Figure 7.3: Differences between average adult male and female faces

ful for biological survival under these unique circumstances. Since each of us alive today is the product of a long unbroken line of successful reproducers, we should still possess much of the functional design that made our ancestors so biologically successful. Viewed from this perspective, our physical and mental attributes become much more understandable.

The modern adult male face and body have many of the characteristics that would have been useful during his long ancestral history as a hunter. The enlarged openings of the mouth and nostrils provide effective passageways for the rapid transport of air to and from the lungs. This enhanced air flow, together with the larger vital capacity of his lungs, is necessary for an adequate supply of oxygen to support the higher metabolic rate and hemoglobin level required for the efficient use of his larger muscle mass. These adaptations are clearly advantageous for a hunting lifestyle. Less obvious are the adaptations surrounding the eyes. Hunting requires a high degree of energy expenditure and inevitably involves profuse sweating from the brow and other regions of the body in order to regulate body temperature. Large bushy eyebrows set close to the eye on a protruding brow ridge provide an effective method for excluding sweat from the eye sockets as well as providing protection from an overhead sun. To be effective, however, a hunter's body requires a hunter's mind.

MENTAL ADAPTATIONS AT PUBERTY

A man's mental adaptations are even more remarkable than the adaptations of his body.[1] Numerous studies have revealed that the male brain excels in visuo-spatial tasks, such as three-dimensional mental rotations and computing the parabolic trajectories of objects in a three-dimensional space. These legacies from the past are still highly prized male attributes and are now an inherent part of sports activities, particularly those that are almost exclusively a male domain. Men who can throw a projectile through space with great accuracy, such as Joe Montana and Michael Jordan, are held in very high esteem and can easily secure a six- or seven-figure income for periodic demonstrations of their prowess. The curious aliens who have been examining us throughout this book, having no knowledge of human evolutionary history, would be greatly surprised by our passionate support for such idols, particularly when it would appear to be more rational to favor those engaged in such practical pursuits as finding a cure for AIDS, reducing crime, teaching, or enhancing child immunization programs. But humans are not rational, at least in the narrow sense of the word. Our psychological adaptations evolved to enhance gene survival in ancestral environments. Men's passions are easily aroused by the sight of a small pack of "hunters" who can skillfully corral an evasive runner with a "pigskin," while their favorite "Lions," "Bears," or "Bengal Tigers" make down after down on a Sunday afternoon. Testosterone levels soar, and the exuberant cries of "Get him!" or "Kill him!" from the excited onlookers attest to the nature of these heightened feelings; they also attest to their evolutionary origins. At the end of it all there are only the high-testosterone winners and the low-testosterone losers. We may ask whether our surrogate victors still reap their reproductive spoils—I suspect they do!

Puberty marks the onset of reproductive competence, a domain where we find major differences between men and women. Unlike men, who produce hundreds of sperm every second, women invest their resources in the production of high-quality ova, one every month throughout their reproductive years. Indeed, the production of large, energy-rich ova is the defining characteristic of females across species. Like men, the female's hedonic amplifiers also respond to hormones, and her emotions rise and fall with the ebb and flow of each menstrual cycle. For many women each ovulation is a time of elevated mood, and premenstrual depression follows each failure to reproduce. The stakes are high, and a woman's feelings wax and wane as each reproductive opportunity comes and goes. Like all feelings, these changes in emotions amplify the potential increases or decreases in her reproductive success. The elevated midcycle mood is a time of potential pregnancy, and the sadness or depression at menstruation amplifies the reproductive consequences of each missed opportunity.

If fertilization does occur, the resulting zygote implants in the uterine wall, where it remains during a long prenatal period. For nine months the new mother provides a continuous supply of nutrients for the developing fetus. Following birth, this maternal investment continues in the form of breast milk, a rich source of nutrients mixed with antibacterial and antiviral agents. Human females, compared with males, clearly make an enormous parental investment in their children. Moreover, this reproductive strategy clearly ensures that a random half of her genes will always be present in her offspring—an outcome that can never be certain for a male. No biological law states that the female must make such a large parental investment; this situation has arisen because, in our ancestral past, any small change to the structure of a female's body that enhanced the survival of her children, relative to females who lacked this modification, would inevitably have resulted in a greater representation of her genes in subsequent generations. Similarly, structural changes that reduced the survival of a female's offspring would decrease the presence of her genes in future generations. As a consequence, genes contributing to adaptive designs increase over generations, while those underlying maladaptive modifications are reduced. To the extent that genes contribute to structural modifications—and they invariably do—females evolve bodies and brains that enhance the survival of their children.

The process that generates adaptive structural modifications can also produce adaptive behavioral designs. An ancestral mother who nurtured her offspring would, on average, leave more surviving children than one who neglected her infants. From a gene survival perspective, it is of little concern whether such nurturing was part of the mother's inherent design, or was developed as a consequence of early exposure to her infants, or was learned over a prolonged period of interacting with her children. Under any of these conditions, a nurturing mother would leave more of her genes than a nonnurturing mother. In a complex changing environment, a mother who learned to care for her children could be as biologically successful, or even more so, than a mother who lacked such flexibility. But if such learning is based on a constrained learning mechanism (similar to the one outlined in Chapter 5), then the emotional value system that underlies nurturing will be transmitted to future generations. Future mothers will then benefit from the flexible design that made their ancestors such successful reproducers.

The major advantage of a learned strategy over a nonlearned strategy is not its independence from genetics. Indeed, a flexible learning mechanism almost certainly requires more genetic design than a rigid program. The advantage of learning lies, rather, in its ability to permit adaptive modifications of behavior in complex changing environments. Learning provides the potential for refining or enhancing adaptive design when environmental conditions are variable; it also, however, harbors the potential for acquiring maladaptive behaviors. To be favored by natural selection, it is only necessary that learning is, on average, more adaptive than other potential alternatives. This is certainly the case here.

If a male could be absolutely certain that his genes fertilized a female's ovum, then he also could be sure that her offspring possessed a random half of his genes. But when fertilization occurs inside a female, no male can ever be totally confident of a child's paternity. Worldwide studies of blood groups indicate that in about ten percent of cases, the man who believes that he is the father of a child is not the biological father. More modern techniques, such as DNA fingerprinting, indicate that the presumed father of the child is not the real father in about fifteen percent of disputed paternity cases. Such studies indicate that even in humans, certainty of paternity is never absolute.

Across species male animals have devised a number of physical and behavioral adaptations to increase the probability that a female's offspring contains their genes. In many insects the male, after the release of sperm, secretes a fluid that rapidly coagulates and serves as a plug that prevents sperm leakage and hinders the introduction of another male's sperm. Some male flies leave their penis behind as a mating plug, and during copulation male mosquitoes reduce the receptivity of females to future partners by introducing a drug, matrone, as part of their seminal fluid. For many male insects penile hooks, spines, and claspers are effective tools for preventing another male from dislodging them during copulation. Also, in species where females have an opportunity to mate with several males in succession, males commonly compete over fertilization by producing a large ejaculate.

The testes-to-body-weight ratio provides an index of the relative importance of the degree of sperm competition across species. In highly promiscuous animals, like chimpanzees, the testes account for 0.27 percent of body weight, whereas they are only 0.02 percent of body weight in gorillas, where promiscuity is quite rare. A human male's testes are about 0.08 percent of his body weight, suggesting some intermediate degree of sperm competition during our evolutionary past. In addition to these biological adaptations, a number of other learned adaptations appear to serve a similar function. Historically, human males have introduced cultural innovations such as chastity belts, female "circumcision," and "honeymoons," in an apparent attempt to decrease the probability of cuckoldry. Traditionally a honeymoon was designed to last for a complete menstrual or lunar cycle, and the bride and groom were encouraged to drink mead, an alcoholic beverage made from fermented honey that was thought to have aphrodisiac properties. Both the "honey" and the "moon" increase the probability of fertilization during this period of exclusive sexual contact.

In his book *Sperm Wars,* Robin Baker proposes that men and women use a number of strategies to affect the probability of fertilization during sexual intercourse. For men, increasing the chances of their sperm fertilizing the ovum and preventing fertilization by other males has always been an adaptive strategy. As a consequence, a normal male ejaculate of 200 million sperm contains about 20 million fertilizing sperm, together with a large number of blockers and killers. According to Baker, blocking sperm enter the cervix immediately after the fertilizers, and because of their unusual shapes they can plug up the pores in the cervical mucus and hinder the progress of any future sperm introduced by another male. Killer sperm, designed to poison the sperm of competitors, offer a second line of defense. The effectiveness of a man's ejaculate is further enhanced by introducing fertilizing sperm that are at different stages of maturity. This arrangement permits mature sperm to advance toward a potential egg in the fallopian tubes while immature sperm hide in cervical crypts, become active at a later date, and can fertilize an egg that is released up to five days after insemination.

In the course of sex with a regular partner, which averages around three times a week, these sperm tactics are effective measures for increasing the certainty of paternity. During sex with a strange woman, however, a man's strategy is quite different. In this case the sperm in an ejaculate can exceed 600 million, with a much-higher-than-normal percentage of fertilizing and killer sperm. This variation in the size and composition of the ejaculate increases the likelihood of overcoming any defenses that have been put in place by a previous male partner. The male's emotional reactions appears to orchestrate this change in tactics. A high level of emotional arousal increases the number and intensity of ejaculatory spurts mediated by the ejaculatory center in the spinal cord. This center is under the control of the hypothalamus, the master controller of the physiological component of emotional reactions. Like other feelings, the intensity of the orgasm reflects the potentially large reproductive gains that could result from this serendipitous encounter with a novel woman, and the hypothalamus adaptively adjusts the man's physiological responses to these fortunate circumstances.

In the competition between males over paternity, women are not passive agents. Under normal circumstances it is the female who decides when, where, and with whom sex will occur. For a female, selecting an appropriate mate is of paramount concern, because acquiring good genes and/or obtaining adequate resources for her offspring have consistently influenced the reproductive success of females over the course of evolutionary history. Errors can be very costly, so the quality of a male's genes, his potential or actual resources, and his commitment to the relationship are the major factors in a female's mate-selection strategy. As we will see, these attributes of males have the greatest influence on a female's emotional attraction. Sometimes, however, it may be in a female's reproductive interest to conceal the identity of the biological father of her child. If, for example, one male possessed resources but a different male had more desirable genes, it could be in her interest to generate a sperm conflict and then influence the outcome toward the male with the better genes.

As with males, such female tactics are not conscious decisions but are orchestrated by emotions that are sensitive to opportunities for enhancing reproductive success. Through mate choice, the timing of sexual contact, and orgasms, a woman can bias the probably of fertilization in favor of her most desirable male. Cervical mucus is most permeable at ovulation, and it is at this time that female-initiated sexuality is at its highest. In addition, female orgasms empty the cervical crypts, stretch the cervical mucus, and increase the number of sperm retained from an ejaculate. That is, female orgasms can decrease any sperm introduced by a regular male partner and increase the sperm introduced by a "good genes" partner. As with a male, it is a female's emotional reactions that bias the sperm competition in favor of the "good genes" lover.

Sperm competitions are probably more prevalent than most of us suspect. Indeed, Baker has estimated that since the year 1900 every single one of us has had an ancestor who was conceived by sperm warfare. Our disbelief is probably based on the common fallacy that males are more promiscuous than females. In fact, every single time a male acquires a new sexual partner, the female involved is also experiencing a new sexual partner. On average, therefore, both sexes are equally promiscuous.

Although men and women do not differ in their average number of sexual partners, the number of partners varies much more within a population of males than within a population of females. That is, a small number of males may have hundreds of sexual partners, whereas many others may die as virgins. In contrast, most females have a small number of male partners, so the difference between females is relatively small. Unlike men, increasing the number of partners does not have a large impact on the reproductive success of a woman. For a woman, the quality of male partners is much more important than their quantity. What circumstances have led to these major differences in sexual strategies?

For both males and females, gene survival depends upon living to reproductive age, reproducing, and ensuring the survival of their children. A multitude of complex subgoals fall within the purview of this master strategy. Overcoming infections, maintaining body temperature, acquiring adequate resources, nurturing offspring, and finding a mate are but a few of the many methods by which both sexes can increase their biological fitness. Although they have many goals in common, the reproductive interests of males and females become dramatically different after puberty. For an ancestral female, bonding to a healthy mate who could acquire resources, and was willing to invest those resources in her offspring, would always have been a successful strategy. This was not the case for males. Imagine an ancestral male who bonded to a single female and lavished his time, effort, and resources on her, irrespective of her sexual conduct or reproductive capability. If such a female was infertile, or her offspring contained the genes of another male, then this "devoted" male would make no contribution to the design of future generations. From a female's perspective, such a male might appear to be a highly desirable mate, since her offspring would enjoy the benefits of his large investment; biologically, however, such a male would be a genetic dead end. Now consider a second ancestral male who placed a high premium on female sexual fidelity and selectively invested his resources in females who showed signs of fertility. These "proprietary" males would have enjoyed more reproductive success than their "devoted" comrades. So we should expect modern males to exhibit the characteristics of their successful "proprietary" forefathers. That is, we should expect modern males to be attracted to females who are both sexually coy and who exhibit reliable cues that are indicative of high fertility.

These differences in the mate-selection strategies of men and women are reflected in a Valentine "chain letter" that I received recently by e-mail.

John Blanchard stood up from the bench, straightened his army uniform, and studied the crowd of people making their way through Grand Central Station. He looked for the girl whose heart he knew, but whose face he didn't—the girl with the rose. His interest in her had begun thirteen months before in a Florida library. Taking a book off the shelf, he found himself intrigued, not with the words of the book, but with the notes penciled in the margin. The soft handwriting reflected a thoughtful soul and insightful mind. In the front of the book, he discovered the previous owner's name, Miss Hollis Maynell. With time and effort he located her address. She lived in New York City. He wrote her a letter introducing himself and inviting her to correspond. The next day he was shipped overseas for service in World War II.

During the next year and one month the two grew to know each other through the mail. Each letter was a seed falling on a fertile heart. A romance was budding. Blanchard requested a photograph, but she refused. She felt that if he really cared, it wouldn't matter what she looked like. When the day finally came for him to return from Europe, they scheduled their first meeting—7:00 P.M. at the Grand Central Station in New York.

"You'll recognize me," she wrote, "by the red rose I'll be wearing on my lapel." So at 7:00 he was in the station looking for a girl whose heart he loved but whose face he'd never seen.

I'll let Mr. Blanchard tell you what happened:

"A young woman was coming toward me, her figure long and slim. Her blond hair lay back in curls from her delicate ears; her eyes were blue as flowers. Her lips and chin had a gentle firmness, and in her pale green suit she was like springtime come alive. I started toward her, entirely forgetting to notice that she was not wearing a rose. As I moved, a small, provocative smile curved her lips. 'Going my way, sailor?' she murmured. Almost uncontrollably I made one step closer to her, and then I saw Hollis Maynell. She was standing almost directly behind the girl. A woman well past forty, she had graying hair tucked under a worn hat. She was more than plump, and her thick-ankled feet were thrust into low-heeled shoes. The girl in the green suit was walking quickly away. I felt as though I was split in two, so keen was my desire to follow her, yet so deep was my longing for the woman whose spirit had truly companioned me and upheld my own.

"And there she stood. Her pale, plump face was gentle and sensible, her gray eyes had a warm and kindly twinkle. I did not hesitate. My fingers gripped the small worn blue leather copy of the book that was to identify me to her. This would not be love, but it would be something precious, something perhaps even better than love, a friendship for which I had been and must ever be grateful. I squared my shoulders and saluted and held out the book to the woman, even though while I spoke, I felt choked by the bitterness of my disappointment. 'I'm Lieutenant John Blanchard, and you must be Miss Maynell. I am so glad you could meet me; may I take you to dinner?'

"The woman's face broadened into a tolerant smile. 'I don't know what this is about, son,' she answered, 'but the young lady in the green suit who just went by, she begged me to wear this rose on my coat. And she said if you were to ask me out to dinner, I should go and tell you that she is waiting for you in the big restaurant across the street. She said it was some kind of test!' "

This story explores the complexity of mate selection and the many different factors that influence the decisions of men and women. In this case it was successful for both parties. The soldier, John Blanchard, ended up with a "young," "beautiful," and "thoughtful" woman, while Miss Maynell found a young, healthy, reliable, and committed mate who was not likely to be distracted by other young women. Both parties obtained exactly what they wanted, but their goals and tactics were quite different.

John initiated the relationship because he believed that Miss Maynell's "soft" handwriting reflected a thoughtful soul and an insightful mind, but when the relationship developed into "a romance," he soon requested a "photograph." Having revealed his interest to be of a physical nature, Miss Maynell quickly took control of the romance and orchestrated the circumstances under which they would meet. Miss Maynell already knew that John Blanchard was healthy and had a steady income; after all, he was a soldier. (This is an important aspect of the plot because if he had been a garbage man or an unemployed waiter, John Blanchard would have been far less attractive to Miss Maynell.) In addition to these attributes, John had also demonstrated his commitment by maintaining an active interest in Miss Maynell over a long period of time. Indeed, insisting on a long courtship prior to a sexual relationship is an important element of female mate selection.

For John, however, there was one last hurdle to overcome, the fidelity test, and it is the structure of this test that makes the story so charming. For John, the fidelity test was difficult. He was forced to make a choice between a woman he had met just a few seconds earlier and a second woman with whom he had developed a strong nonphysical relationship over a long period of time. It doesn't appear like a difficult decision, but the crucial point was the difference in age and beauty between the females. Why should these factors be so important to a human male?

David Buss, author of *The Evolution of Desire,* has examined the factors that contribute to male and female mate preferences across many different cultures.[2] These studies confirm the theoretical predictions arising from an evolutionary viewpoint as captured by the Blanchard–Maynell story. In general, women are attracted to healthy, committed males who have, or have the ability to acquire, resources. In contrast, males appear to be attracted to coy and "beautiful" females in their reproductive years. These observations, however, raise further questions. What is it about the appearance of a woman that leads us to call her beautiful? What is beauty?

IN THE BRAIN OF THE BEHOLDER?

Physical beauty does not exist in the external world. Rather, it is a positive feeling that is evoked by a configuration, often a complex configuration, of visual, auditory, and other sensory components. For a hen, there is probably nothing more beautiful than a large clutch of eggs. Our persistent aliens might view the human body as ugly and repugnant, but humans often perceive the appearance of the opposite sex as beautiful or handsome. I have been arguing that feelings act as discriminant hedonic amplifiers that exaggerate the reproductive consequences of certain physical and social events. If this is the case, then what is the biological importance of a feeling like beauty? What physical attributes are responsible for eliciting this feeling? Examining the factors that contribute to female beauty, or male handsomeness, offers a unique opportunity for evaluating the proposed theory of human feelings.

The human notion of physical beauty can be defined as the positive feelings evoked by the physical appearance of an individual, in the absence of any other information. The old adage "Beauty is in the eye of the beholder" attests to the common belief that there are no accepted standards by which this ephemeral property can be measured or judged; that is, we all use different criteria. A second maxim, "Beauty is only skin deep," reflects the common notion that beauty is a superficial and unimportant aspect of attractiveness. Finally, the enduring proverb "Don't judge a book by its cover" cautions us to ignore this trivial attribute when making judgments. This cultural wisdom, however, appears to fall on deaf ears. The pursuit of beauty is, and always has been, an integral part of every known human culture. In the modern world beauty enhancement has spawned a multibillion-dollar worldwide industry, and many otherwise normal individuals spend a great deal of time, effort, cost, and pain attempting to improve their physical appearance. It is difficult to reconcile these behaviors with the pursuit of a trivial attribute that has no important value and cannot be measured, and that we all have been cautioned to ignore. Face-lifts, cosmetics, collagen injections, liposuction, and silicon implants all argue against our common convictions and suggest that idealized standards of beauty do exist, at least within cultural groups.

Beauty is not ignored. Experimental evidence shows that a number of important decisions—such as courtroom judgments and sentencing, hiring decisions, and mate choice—are all influenced by physical attractiveness. Indeed, the entire advertising industry is based on the premise that attractive individuals can be used to entice others to buy commercial products, and the strategy appears to work. This bizarre schism between what we say and what we do raises a number of important questions. Do some idealized standards of human beauty exist either within or between cultures? If so, are these standards arbitrary, or do they reflect some basic and universal biological design? If beauty does have biological value, then what is it? Only recently have experimenters attempted to address some of these interesting issues.

BEAUTY AND NATURAL SELECTION

From an evolutionary perspective, any characteristic of the face or body that enhances biological survival will increase in frequency over generations as a consequence of natural selection. For example, a wide, flat nose is best in the tropics, where it needs to contribute only a small proportion of the required moisture and warming to the already hot and steamy air. Conversely, in the cold and arid climates of the world, a much longer nose is required to condition the air. As a consequence, the average facial features within any breeding population should evolve, over generations, to ultimately reflect the most desirable configuration for biological survival in any local environment. Individuals who are attracted to an average face would enjoy an increase in fitness because their offspring would reap the benefits of all of these successful adaptations. A theory of physical beauty based on natural selection would therefore predict that individuals should be attracted to the average face and body in a breeding population and consider such characteristics to be beautiful.

Although it is counterintuitive, the "average is beautiful" theory has received a great deal of experimental support. It is possible, for example, using image-processing software, to generate an average face by combining the digital photographs of many different faces. In experiments these average faces have been judged to be more physically attractive than the individual faces that make up the average, and furthermore the attractiveness of a multiface composite increases with the number of faces that are used to make the average. There are, however, two interpretations of this outcome. Either the average face in a population is the most attractive face, or the average face is more attractive than most faces but is still not the most attractive face in a population.

A second way to find the most attractive face is to use the FacePrints program described in Chapter 3. Finding a beautiful face in a virtual face-space is almost identical to evolving the face of a criminal suspect. In this case, however, the experimental participants rate composite faces for beauty rather than resemblance to a culprit. These beauty ratings guide the search toward the most attractive face in a virtual face-space. That is, it will "evolve" the face that each participant considers the most attractive. Beginning with a first generation of random faces that are distributed throughout a virtual face-space of more than 17 billion possibilities, the genetic algorithm will converge, over generations, on the particular face that each participant considers to be most beautiful. The features and proportions of these final composites can then be compared with the attributes of average faces in order to uncover any facial characteristics that are correlated with beauty.

My student Melissa Franklin and I instructed male and female volunteers to "evolve" their most beautiful female Caucasian face using the FacePrints software (Figure 7.4).[3] In sharp contrast to conventional wisdom, the characteristics of a beautiful female face were found to be neither average nor arbitrary. Both men and women generated female faces with very similar attributes, and these attractive composites were significantly different from the average female face in the population. The result is shown in Figure 7.5. First, the lower jaw region of attractive female faces was found to be smaller than average. Second, beautiful faces had fuller-than-average lips. These results remained controversial until they were replicated by independent scientists working in several different cultures and using different experimental procedures. It was soon found that faces judged to be attractive in one culture were also deemed attractive by members of other cultures, and these attractive faces possessed the same nonaverage characteristics. It appears

Figure 7.4: Interface of FacePrints computer program

that an attractive female face in any culture can be constructed by finding the average face in that culture and then adjusting the lower jaw length and lip fullness, as shown in Figure 7.5.

   If a beautiful female face differs from the average by virtue of possessing a smaller lower jaw and fuller lips, are these visible cues transmitting some biological message? Some insight into this question can be gained by examining the features of the female body that are also considered to be beautiful. Although preference for the total amount of fat on a female body has varied both historically and across cultures, the desirable distribution of fat, measured by the waist-to-hip ratio, has remained very stable. Researchers have shown that a waist-to-hip ratio of about 0.7 has been considered beautiful across recorded history and across many different cultures.[4] The waist-to-hip ratio has been shown to vary with the relative amounts of free androgens to estrogens in the blood stream. Furthermore it has been shown that deviations from the 0.7 ratio are associated with a marked reduction in fertility. These attractive proportions, correlated with low androgen and high estrogen

Attractive Proportions   Average Proportions

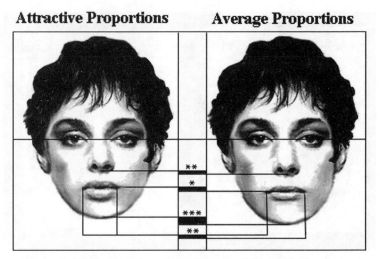

Figure 7.5: The proportions of an average face (right) and the proportions of a beautiful face (left). The only features that vary are the mouth and lower jaw.

levels, may well serve as observable, reliable indicators of high fertility. It is noteworthy, therefore, that the hormonal factors influencing the waist–to–hip ratio are the very same ones that appear to influence facial features and proportions.

Prior to puberty, the lower jaws of male and female faces are similar in length (Figure 7.2). At puberty, however, males undergo a rapid growth spurt controlled by androgens, especially testosterone, accompanied by a rapid increase in the growth of the lower jaw. The female jaw undergoes a similar modification, so a lower-than-average androgen exposure at puberty would lead to the shorter lower jaw found in attractive female faces. Based on the normative growth curves, beautiful female faces have a jaw length that is more than one standard deviation shorter than the average adult female face. Like the jaw, female lips also vary with age and show a marked change in fullness during the years immediately following the onset of puberty. Lip fullness appears to depend on estrogen-controlled fat deposits, since it increases in parallel with the fat deposits on the hips and breasts and reaches its maximum value between fourteen and fifteen years of age. The larger-than-average lip fullness characteristic of beautiful faces may therefore provide an indication of higher-than-average estrogen exposure during this developmental phase. Since the combination of low androgens and high estrogens is associated with an optimal waist-to-hip ratio and high fertility, the combined evidence suggests a general theory of female beauty that applies to both the body and the face. The feelings evoked by a beautiful female face or body appear to be an emotional response to those features and proportions that are reliable cues for female fertility. This is strong support for the theory of feelings developed throughout this book; feelings are evoked by circumstances that have consistently posed a threat or offered a benefit to our biological fitness. Beauty, like all of our other affects and emotions, appears to be an omen of reproductive success.

A natural selection viewpoint, as we have seen, would predict that the average face in a population would be the most beautiful face. Despite its appeal, however, natural selection alone can not account for the consistent experimental finding that very specific nonaverage facial features and proportions are considered to be attractive across cultures. The observation that very attractive faces are systematically different from the average across many different cultures, even when exposure to Western media is minimal, is better explained by the theory of sexual selection.

BEAUTY AND SEXUAL SELECTION

Charles Darwin defined sexual selection as "the advantage that certain individuals have over other individuals of the same sex and species, in exclusive relation to reproduction." That is, members of the same sex may compete with one another over opportunities to breed with members of the opposite sex, and such competitions can lead to the evolution of physical or behavioral adaptations that provide some advantage in these intrasexual competitions. From a gene survival viewpoint, there is really no difference between natural and sexual selection, since any such behavior or structure will be favored over generations only if it contributes to the survival of the genes of those who possess it. But if we are interested in the physical outcome of the evolutionary process, it is useful to distinguish between characteristics that evolved because they enhanced personal survival, and those that evolved as a consequence of intra-sexual competitions over direct access to reproductive opportuni-ties. From a sexual selection viewpoint, a characteristic that is attractive to the opposite sex may enhance reproductive success even though the feature may be detrimental to the individual's per-sonal survival.

This effect can be clearly seen in the case of a peacock's tail. A large colorful tail may be cumbersome and dangerous to a pea-cock's personal survival, but it may have become more elaborate over generations because it served as a signal for an attribute that peahens (females) found highly desirable, probably parasite resis-tance. Under these conditions the tail length would grow over gen-erations until its average benefit (in gained reproductions) was equal to its average cost (in lost reproductions). The average tail in a population of peacocks, then, is a compromise between the influ-ences of these two factors, while the most attractive tails possess the extreme characteristics that are most desirable to the opposite sex. Peahens do indeed prefer longer-tailed males, whose reproductive success is higher as a consequence.[5]

And what do women look for in men? Clearly symmetry— a signal of "good genes"—plays a part. Across species, sexually selected male features are energy-demanding displays that are clearly perceptible, intricate, and bilaterally symmetrical. The pea-cock's tail, for example, is a large, conspicuous, bilaterally symmet-rical display of colorful "eye spots." A conspicuous display such as this could serve as a sensitive indicator of developmental stability. Large developmental instabilities can be recognized from birth defects, but small random deviations from bilateral symmetry are sensitive indicators of even minor disturbances in the expression of genes during development. A large, intricate sexually selected dis-play offers an ideal platform for a male to flaunt his good genes by exhibiting a high degree of symmetry, and for a female to recog-nize developmental instability by detecting asymmetrys.

But does the symmetry of sexually selected displays actually increase the mating success of males? Two biologists, Anders Møller and Randy Thornhill, have addressed this question by analyzing the results of fifty-five studies that looked at the relationship between asymmetries and mating success or sexual attractiveness across forty-two species.[6] The results are compelling. Across species, symmetrical characteristics enhanced both the attractiveness and the mating success of males. In humans the facial symmetry of men has been found both to be attractive to women and to increase a man's sexual opportunities.

BEAUTY AND FERTILITY

Sexual selection is not a one-way street. Indeed, among humans, males are often more concerned about the physical appearance of their potential mates. It may be important for females to detect "good genes" in their prospective mates, but males have an additional concern: fertility. Just as the symmetry of males displays their "good genes," females can flaunt their fertility using other features of their face and body. Indeed, female fertility displays are very common in across species. In primates female signals of fertility have been observed in at least fifty-five different species. Human females are no exception; fertility cues are displayed all over their face and body.

In humans the reproductive advantage of such signals is clear. A woman's potential fertility begins shortly after puberty, reaches its maximum in her early twenties, and then declines rapidly. By the early thirties the female fertility rate has dropped to about eighty-five percent of maximum, and by the early forties it has declined to about thirty-five percent. For a human male, these circumstances create a small window of opportunity, so the ability to detect and respond to fertility signals has always been a successful strategy. Females displaying such cues could attract higher-quality mates and so enhance their reproductive success relative to their rivals. Indeed, the observed trend toward earlier menarche, and the cross-cultural practice of exaggerating fertility cues, may be a consequence of such intrasexual competition.

Sexual selection appears to offer the best explanation for why specific facial factors influence female beauty. The average female face is attractive, but it is a compromise between the influences of natural and sexual selection. Average is attractive, but the most attractive faces are different from the average because they possess certain cues, such as full lips and shorter lower jaw, that are indicative of a higher-than-average fecundity. A larger and stronger lower jaw may be more beneficial for personal survival, but as with the peacock's tail, sexual selection will favor the length that is attractive to the opposite sex. From this perspective we should expect that beautiful faces, across cultures, will vary from the average in a systematic manner. This appears to be the case. Full lips and a shorter-than-average lower jaw have been shown to enhance beauty across many different cultures.

Because a small jaw and full lips are also characteristics of a youthful face, some experts have proposed that youthfulness, or a childlike appearance (neoteny), may underlie our judgments of female beauty. There are really two variations of this hypothesis. One states that a beautiful adult female face has a childlike appearance (the neoteny hypothesis), while the second states that the beautiful adult female face has the characteristics of a young post-pubertal female, about sixteen years of age. The first hypothesis is based on the observation that very young mammals, including human infants, have a large forehead, retracted chin, and relatively large eyes that are located below the midpoint of the face. Humans appear to be attracted to this facial configuration, and these characteristics are common in popular cartoon characters such as Mickey Mouse and Garfield. In addition, an infant's face undoubtedly evokes an emotional response in a viewer. People looking at a baby will smile on almost every occasion. These feelings, however, appear to be nurturing and have no romantic or sexual connotations. We consider a child's face to be "cute" but not beautiful. Indeed, if adult features are scaled to size and arranged in the proportions of an infant's face, the result is quite sinister and even grotesque.

The second hypothesis is more compelling. Many young adult females, about sixteen years of age, do indeed possess the proposed fertility cues: large lips and a shorter lower jaw. It is therefore reasonable to propose that the youngest females who exhibit these cues are considered to be most beautiful. The evidence, however, does not entirely support this interpretation. For example, when independent judges were asked to estimate the age of beautiful faces "evolved" using the FacePrints software, they found them to be in their early twenties. Also, between 1953 and 1990 the average age of *Playboy* centerfold models, selected on the basis of their physical beauty, was 21.3 years. These observations are consistent with the fertility hypothesis since, on average, fertility is low in the years immediately after puberty but increases until it reaches a peak in the early twenties. These findings also suggest that additional cues, perhaps age-related factors, may also contribute to female facial beauty. That is, beauty may be evoked by facial features that are influenced by growth hormone, in addition to the proposed estrogen and androgen effects. If this is the case, then our feelings of beauty are exceptionally well tuned to the age of maximum fertility.

It is possible to distinguish between the youthfulness and fertility views of beauty by conducting further experiments. Using the FacePrints software, volunteers can be instructed to "evolve" either their most attractive female face, or their most youthful adult female face, using the same database of facial features. If youth and beauty are equivalent concepts, then participants evolving a beautiful face would unwittingly be evolving a youthful face as well, and vice versa. That is, if youth and beauty are equivalent, then the evolved faces should be similar; if youth and beauty are different, then the evolved faces should differ. The maximum fertility hypothesis predicts that the evolved beautiful and youthful faces will be different. More specifically, beautiful female faces should possess age-related cues that will cause them to be perceived as being in their early twenties, much older than the evolved youthful faces.

My first attempt to define the differences between youth and beauty was a study conducted on the World Wide Web. During the month of April 1996, a FacePrints website was constructed where visitors from all over the world could participate in evolving a youthful or beautiful female face. Because the rate at which pictures can be downloaded from the web is limited, the FacePrints program required some modifications. The web version began by displaying a first generation of sixteen randomly generated female faces and allowed visitors to the website to enter a beauty rating for each one. After they had entered these ratings, they were then shown the same sixteen random faces and were now asked to rate then for youthfulness. Over the course of the experiment, visitors on even days of the month rated first for beauty and then for youth, whereas visitors on odd days rated first for youth and then for beauty.

When the first generation of faces had been evaluated by more than one hundred people, a second generation was produced, using the FacePrints program. To create a second generation of attractive faces, a mean beauty rating was first calculated for each of the first generation of sixteen. Since each face was stored in the computer as a long binary string, its "genotype," the mean beauty rating of each face provided a measure of the fitness of each genotype. A second generation of faces was then computed by selecting, crossing over, and mutating the first generation of genotypes in proportion to their fitness ratings. The best way to visualize this process is to think of a lottery, with each first-generation genotype having a number of lottery tickets equal to its beauty rating. Now mix the tickets together, and select two at random. These two selected genotypes now cross over and mutate (as shown in Figure 3.2), and the two offspring are then developed into two second-generation faces. Since genotypes receiving a high beauty rating have more tickets in the lottery, they are more likely to be selected; this is selection in proportion to fitness.

This process was repeated seven more times, producing a total of sixteen second-generation faces. They were then uploaded to the website for rating. Exactly the same procedure was used to breed a second generation of youthful faces, which were also uploaded to the website. Over the following days visitors to the website were asked to rate the second generation of attractive faces for beauty, and the second generation of youthful faces for youthfulness. The process continued for many weeks, with a new generation of youthful of beautiful faces being posted whenever sufficient votes had been received for any generation. Starting with identical first generations, the two populations of faces quickly diverged from each other and then converged onto different locations in face-space (Figure 7.6).

Needless to say, the website was very popular and attracted votes from Argentina, Australia, Bahrain, Belgium, Canada, Chile, Denmark, Ecuador, Finland, France, Germany, Greece, Hong

Figure 7.6: Evolved attractive and youthful faces

Kong, Hungary, Iceland, Indonesia, Ireland, Israel, Italy, Japan, Korea (South), Lithuania, Malaysia, Mexico, Netherlands, Norway, New Zealand, Pakistan, Philippines, Portugal, Singapore, Slovak Republic, South Africa, Sweden, Switzerland, Thailand, Turkey, United Arab Emirates, United Kingdom, United States, and Uruguay, with ages ranging from eight to eighty. Over the course of several experiments, the hit rate soared from less than 40 per day to a rate of more than 400 per hour.

During periods of high activity, it was possible to compare the mean beauty ratings of a population of faces from hour to hour. The agreement among the visitors was remarkable. The correlation between beauty ratings every hour (average r = 0.96) was similar to that between ratings of youthfulness (average r = 0.95). That is, visitors to the website were just as consistent in their estimates of beauty as they were in their estimates of the ages of female faces. Given the diversity of the population who voted, this is an extraordinary consensus. Despite the common belief that beauty is "in the eye of the beholder," there appears to be a great deal of agreement among us.

The study indicates that youth and beauty are indeed different, but a more controlled study is required in order to ascertain the nature of these differences. In a second study a group of seventy students were required to evolve either their most beautiful or their most youthful Japanese female face, using the FacePrints software. Japanese faces were used in order to determine if the same cues were used for judging the age and beauty of faces from different cultures. After all seventy faces were generated, they were mixed with forty control pictures of young Japanese women, and all 110 faces were then rated for age and beauty by eighty independent judges. This procedure allowed the youthful and beautiful faces to be compared to control faces as well as to each other.

Both the youthful and the beautiful faces were rated as higher in beauty than the average control faces. Specifically, both groups of faces had significantly larger lips and eyes, a narrower nose, and a shorter and narrower jaw than the control group. But only the youthful group was perceived to be significantly younger than both the beautiful and the control faces. The relationship between perceived age and beauty is shown in Figure 7.7.

Consistent with our prior findings on Caucasian faces, the beauty of a Japanese female face was found to be maximum when its perceived age was about 22.4 years of age. The bottom graph in Figure 7.7 shows that the beauty of the youthful faces increased with perceived age, whereas the perceived beauty of the beautiful group was at a maximum in the early twenties—close to the age of maximum fertility. Finally, significant age cues distinguished the youthful and beautiful faces. Faces that were rated as very beautiful possessed a longer eye-nose distance, slightly smaller eyes, and a wider nose and jaw than the youthful faces. These differences can be seen in Figure 7.8.

The fertility hypothesis proposes that the feelings elicited by a female face or body are a function of specific physical cues modulated by pubertal hormone levels. Figure 7.9 shows a female Caucasian face that possesses many of these desirable features and proportions. These cues may serve as reliable indicators of fertility and are found to be attractive without any awareness of their relationship to reproductive success. Beauty, like other conscious feelings, appears to act as a filter (or discriminant hedonic amplifier) that exaggerates the reproductive consequences of environmental cues associated with relatively minor fluctuations in reproductive success. Although these cues appear to be the most important factors regulating our assessment of beauty across cultures, extreme

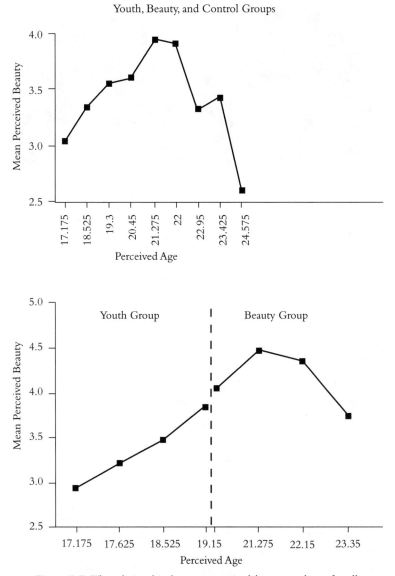

Figure 7.7: *The relationship between perceived beauty and age for all faces (top) and for youth and beauty groups separately (bottom)*

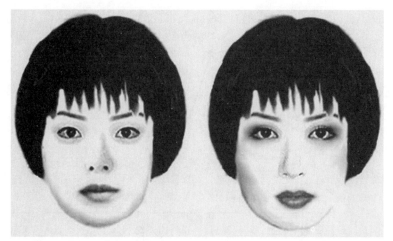

Figure 7.8: Differences between a youthful (left) and a beautiful (right) Japanese female face

degrees of attractiveness may depend upon additional learned factors that are specific to individuals. If learning is an evolved ability that permits individuals to modify their behavior within their lifetime and further refine their biological design, it likely also plays a role in refining our concept of physical attractiveness. Ultimately a comprehensive theory of beauty should integrate learned factors with the other evolutionary factors that appear to shape our preferences for the physical appearance of the opposite sex.

DESCRIBING IS NOT PRESCRIBING

Research examining the origins of human feelings is often misunderstood, but one misconception is so prevalent that it is deserves its own name: the naturalist's fallacy. This fallacy is the notion that since an attribute like beauty arises from our biological nature, it must be "good" or even desirable. Much of our biological design, however, arose in societies that are very different from modern industrialized society. A good example of this is our biological preference for the taste of sugar. In our ancestral environment the major source of sugars was ripe fruit, and those who liked the taste of sugar would inevitably benefit from a highly nutritional diet. In the

Figure 7.9: Female Caucasian face with maximum fertility cues

modern world refineries have separated the sugars from the nutrient so that what was formerly an adaptive "sweet tooth" is now a dangerous preference. In a similar manner a preference for a body or face that indicated reproductive competence was certainly important for our forefathers, but in an era of fertility drugs and contraception, that preference may have little relevance. Nevertheless, like sugar, our "sweet tooth" for beauty is still a part of our nature and probably will remain so into the foreseeable future. Understanding the origins of our preferences—for sugar, for beauty—and their relationship to reproductive success is very different from stating that they are "good" or "bad." Our emotional inheritance is neither "good" nor "bad"; it just is. It does, however, have a profound effect on the way we reason about the world around us.

EIGHT

❧

Legacy or Logic?

MOST OF US BELIEVE THAT WE LIVE IN A WORLD THAT has three spatial dimensions and a strange fourth dimension that we call time. Objects that are close to us in space elicit conscious experiences in the form of clear visual pictures, whereas more distant objects are perceived as blurred or fainter images. (In fact, we really construct our image of distant objects from time-delayed two-dimensional patterns on the surface of our retinas, but since light travels very fast, the delay is negligible for earthbound objects.) We can travel through the three spatial dimensions in any direction we choose, but we can't move back and forth through time. Indeed, everything in our three-dimensional space appears to be forced to move through time together, and all such movements are confined to a single direction: from past to future. Objects that are separated from us in time are not represented as visual images at all. We perceive our past as memories, and we can "see" only a short distance into the future using our ability to reason. Memory and reasoning appear to be our evolved representations for locating events in the temporal dimension.

Like distant events in space, incidents in our past are remembered less clearly than recent happenings, and future events are very fuzzy indeed. It is possible to simulate the process of seeing the past as visual images by pointing a television camera into its own monitor and then locating your hand between them. In the monitor you can now view events in time as a series of pictures that form a long tunnel running off into the distant past. If you move your index finger, the front image will show its current position, but as you look farther down the tunnel, you can see a trail that reaches back into the past and displays the earlier positions of your finger. Why did we not evolve such a photographic or cinemagraphic memory that would allow us to form pictorial representations of our past? After all, every moment in time is already perceived as a three-dimensional visual image, so it would appear easier to "store" these sequential pictures than evolve an entirely different abstract representation of our past. And what about the future—could it also be represented visually? The movement of simple objects through space and time can often be predicted with a high degree of accuracy. Computers can "see" where a missile will land and even predict the damage that it will inflict when it reaches its destination. Could we have evolved a visual imagination to clearly represent the future consequences of our actions? The problem here, of course, is that the future depends upon the interaction of many, many factors, and most of them are unknown to us.

Our limited ability to recall the past and predict the future may be a consequence of some physical constraint that sets an upper boundary on the amount of information that a biological brain can possibly store. Or perhaps selection pressure has been insufficient because a clear view of the past and future would add little to our ability to survive and reproduce on this planet. With respect to seeing into the past, the first hypothesis is certainly true. Even the most sophisticated computers are taxed to their limit when they are required to store long sequences of visual images. This problem could be greatly reduced, however, by storing images only of events that are directly related to our biological survival. As we have seen, this is exactly what we do whenever we learn—only biologically relevant events, defined by our feelings, initiate the formation of memories. But we don't really need a visual image of the past that would allow us, for example, to count the number of rocks on a remembered garden wall or see the entire image of a book that we have just read. We need only store the meaning of past events in such a form that we can retrieve this information to solve current problems, or use it to predict the future. This is done in our compact semantic memory. We still don't know exactly how it works, but internal and external contextual cues are certainly important for storage and retrieval, and as discussed in Chapter 5, every associative memory is stored along with its positive or negative consequences. It is true, of course, that we sometimes forget things, but our most biologically important memories—those associated with intense feelings—appear to be saved in a much more permanent form. Our semantic long-term memory may be imperfect, but it is certainly adaptive. When it is lost—in Alzheimer's disease, for example—we suddenly become aware of just how important it really is.

What about our view of the future? Would it be possible or useful to see a picture of the world at some fixed interval into the future? Since we already construct a current image of our world, a future image would appear to be no more expensive in a biological sense. The problem here, of course, is not the image itself but the possible accuracy of such an image. For physical phenomena, such as how the world would look if we walked to the other side of the room, a photographic representation could be quite accurate. In fact, computers can already generate images of rooms from any point of view, so such representations would appear to be quite feasible. The problems become more complex, however, when objects in the world are moving, and they become insurmountably difficult when we consider the social world. Tomorrow's weather may be predictable to some degree, but try predicting the actions of another human being! *Combinatorial explosion* is a good term for describing the enormous difficulty of representing all possible events that could arise in the social domain. It requires only a little thought to conclude that social events are the most difficult to predict; they are also, however, the most biologically advantageous. Anyone who could accurately predict the behavior of others would have an enormous reproductive advantage. To put it another way, natural selection should always have favored social reasoning ability, but pictorial representations may not be the best solution for achieving this goal.

As with memory, we really don't need to predict every facet of the future; we need only represent those aspects that are of potential biological importance. Second, we should not expect our reasoning to "see" into the distant future; social interactions require moment-to-moment predictions that are constantly updated according to the responses of others. Third, any ability to predict the future is possible only by using our memory of the past. Indeed, the only functional usefulness of memory lies in its ability to provide information that is relevant to current or future situations. But memories of the past do not take the form of pictures; they are semantic memories, in the form of stored hedonic outcomes, indexed by environmental cues and biologically relevant feeling states (Figure 5.1). This format, however, is ideal for reasoning.

THE NEED FOR SOCIAL REASONING SKILLS

Reasoning depends upon the ability to simulate a variety of alternative scenarios that are possible in the current situation and to make decisions based upon their probability of occurrence and their expected hedonic consequences. By selecting behaviors that are most likely to lead to the largest positive hedonic tone, the alternative that feels best, a decision-maker is unwittingly selecting the outcome that has the highest probability of increasing his or her reproductive success. Social reasoning, however, has an additional complexity that is not inherent in the inanimate world. Other individuals besides ourselves have their own unique hedonic outcomes that influence their decisions. This presents an enormous problem since it is not possible to examine the feelings generated inside the brain of another human being. As a consequence, accurate predictions of social behavior become almost impossible.

The apparent solution is to represent the feelings of others without actually feeling them. Presumably this "mind reading" is possible only because we can recall the stored hedonic outcomes that we have personally experienced under similar conditions and can improve on this model through our interactions with specific individuals. Our stored knowledge of hedonic consequences makes it possible to reason "If I do this, then he will probably do that, and I will feel good when he does that!" We have entered the world of "self-consciousness," a virtual reality where actions and reactions of self and others are simulated and evaluated using hedonic outcomes, and the probabilities of those outcomes.

This social reasoning ability is certainly one of the most complex and least understood attributes of our brain; it is also a skill with enormous reproductive consequences. Indeed, the evolution of social reasoning skills may have been the major impetus for the rapid evolutionary expansion of the human brain. That is, our big brain may have evolved, not as a tool for conducting differential calculus, but as a tool for coping with the complexities of social decision-making. The frontal lobes, in particular, appear to be critically involved, since social decision-making is most severely disrupted by frontal lesions. In his book *Decartes' Error,* Antonio Damasio documents the severe social disturbances that follow frontal lobe damage.[1]

Because emotions are so closely tied to events in the social world, our context-addressable memory of stored hedonic outcomes is primarily a product of our social interactions. Such memories should therefore be better suited for predicting social outcomes than for any other type of reasoning. Our reasoning skills should be particularly effective in the social domain and perhaps somewhat less effective in more abstract or numeric domains.

The Prisoner's Dilemma game is a useful tool for examining how we reason in social situations. The game gets its name from a common situation that confronts two prisoners who have both been accused of jointly committing a crime. They are both guilty, but neither has admitted it. The prisoners, detained in separate cells, have to decide whether to cooperate in maintaining their innocence, and the outcome for each prisoner depends on both his own decision and on the decision of his accomplice. The game is thought to represent many social decisions in the real world, and the payoff matrix shows the social dilemma.

The payoffs range from a light sentence (four points) to a heavy sentence (one point). If both prisoners cooperate (C), they both win three points, and if they both defect (D), they get two points each (Figure 8.1). Both outcomes translate into moderate

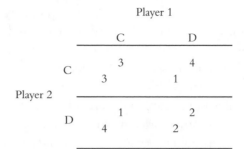

Figure 8.1: *Payoff matrix for Prisoner's Dilemma game. The numerical values show the payoff that each player receives according to whether the other player cooperates (C) or defects (D).*

jail sentences. A twist, however, really complicates matters. If only one prisoner defects, by admitting that they are both guilty of the crime, then the defector gets four points (a light sentence) but his accomplice will get only one (a long jail sentence).

Imagine you are required to play this game, just one time, with a total stranger. What would you do? If you defect and the stranger cooperates, then you will win four points, the best score that you can get. However, if the stranger also defects, you will win only two points. This second outcome is less than you might have hoped for, but remember that if you had cooperated under these very same conditions, when the other player defected, then you would be worse off, with only one point. It's easy to see that when you are going to play this game on one single occasion, it's always best to defect. A real predicament arises, however, when you are destined to play the game many times, with the same individual, over the course of a lifetime. Remember, this game is designed to be a model of real-life social decisions, and points can represent any resource that is valuable to both players. If both players reason as outlined above, then defection will be constant, and each will win two points on every occasion. Notice, however, that if both players had coop-erated they could have done much better, gaining three points each on every occasion. This is the real dilemma in the game.

THE ROLE OF FEELINGS IN SOCIAL REASONING

Social psychologists, political scientists, and mathematicians have puzzled over and written about almost every possible strategy for playing and winning this intriguing game. In recent years computer tournaments have pitted strategies against each other in an attempt to uncover which would be most successful in the context of many different competitors. Surprisingly, Tit for Tat, a very simple strategy, has won in repeated tournaments and is now generally believed to be the optimal strategy for playing and winning this social game.

Tit for Tat always begins with a cooperative move, and thereafter it copies the play that its opponent used in the previous round of the game. It is described as "nice," because it is never the first to defect, and "forgiving," because it will immediately stop defecting as soon as its opponent begins to cooperate. In fact, Tit for Tat never wins an individual game, since its maximum score is always equal to, or a few points less than, the score of its opponent. It is an overall winner, however, because less "nice" strategies have to play against each other, and in these games they do very poorly. It is also true that once a population of players all adopt a Tit for Tat strategy, then no other fixed strategy can ever do better.

These observations raise a number of interesting questions. Do humans use a Tit for Tat strategy? How could it arise within a population that initially contained many different strategies? How would it be implemented within an individual: as an automatic mechanism, a consciously worked-out strategy, or simply a feeling-based decision mechanism? Some of these questions were examined in two computer tournaments that were part of graduate seminars held in the psychology department at New Mexico State University in 1994 and 1996, and simulated feelings were the key to a successful strategy.

The rules for playing in these Prisoner's Dilemma tournaments were somewhat different from the rules used in previous tournaments. First of all, only nondeterministic programs were allowed to participate. In prior tournaments the submitted programs always obeyed a set of programmed rules. That is, anyone knowing the program code could predict the behavior of any contestant, because the programs were based on fixed strategies. In the graduate seminars, however, students were required to write programs that "evolved" rules using a genetic algorithm. Using feedback from success or failure over the course of a game, each program was required to generate its own rules for playing the game, in an attempt to maximize its winnings. As we have seen, genetic algorithms employ random crossovers and mutations, so it is impossible to predict the exact rules that will evolve. It was reasoned, however, that if Tit for Tat was really the best strategy for playing and winning the Prisoner's Dilemma game, then such innovative programs should evolve Tit for Tat rules over the course of a game.

Evolving strategies for playing the game can be implemented by using a classifier system. This procedure begins with a set of rules, called classifiers, that are written as binary strings and can therefore evolve using the standard genetic algorithm. For example, a rule such as (11100001) would mean that if your opponent cooperated three times (111) and then defected four times (0000) during the last seven plays, then you should cooperate (1). If applying this rule leads to a favorable outcome, such as winning three points, then this rule increases in fitness. Now imagine that a program begins with thirty-six randomly generated classifiers that gain or lose fitness according to the outcomes they produce. The rule used on any trial is selected on the basis that it best fits its opponent's behavior pattern over the previous seven trials and has the highest fitness. The generality of rules can be further increased by using an asterisk ($\star$) to specify a "don't care" option, meaning it does not matter whether the opponent chose to cooperate or defect. That is, a rule such as ($\star\star\star\star\star\star$11) would specify a cooperative move (the last 1) if an opponent cooperated on the last trial (indicated by the penultimate 1), irrespective of their prior behavior (simply notated by asterisks); a rule like ($\star\star\star\star\star\star$00) would mean "defect" if your opponent defected on the last trial. Indeed, these two rules together define the Tit for Tat strategy. In a classifier system, however, new rules can evolve by crossing over and mutating the most successful rules, using the standard genetic algorithm described in Chapter 3. If Tit for Tat is the best strategy, then we should expect the Tit for Tat rules to evolve over time.

The first competition held in 1994 played seven nondeterministic classifier systems and a single deterministic strategy, Tit for Tat, in a tournament where every player had to play every other player (a Round Robin tournament). Within this group of creative strategic players, Tit for Tat did not fare well. The winner was a classifier system that played the first twenty plays using random classifiers but then rapidly evolved new rules using its genetic algorithm. This program did reasonably well playing against Tit for Tat, but its real advantage lay in outwitting weaker opponents. In contrast, Tit for Tat was unable to exploit these weaker opponents and therefore achieved a lower overall score. Despite its success, however, the top strategy was still easily defeated by a human player. Its weakness against a human opponent lay in its inability to quickly switch strategies after a long series of cooperative moves. Repeated cooperation could "lull the program" into a cooperative "state of mind," and now a sudden series of defections by the human did not meet with instant retribution. Something was missing in this program: it didn't get "angry." For the 1996 tournament, the program was fitted with simulated feelings!

STRATEGY:	Feel	Guts	Blood	TiTa	Risk	Sweat	Power	Tears	Rigid
TOTALS:	4187	4169	4134	4045	3987	3844	3843	3704	652

Figure 8.2: Results of the 1996 tournament. Each player was matched against every other player, in a Round Robin tournament.

The results of the 1996 tournament are shown in Figure 8.2. Tit for Tat (TiTa), the winner of the prior competition (Power), a classifier system without a genetic algorithm (Rigid), five new opponents (Guts, Risk, Blood, Sweat, and Tears), and the new version of Power with simulated emotions (Feel) were all entered into a second Round Robin tournament. All the new programs (Guts, Risk, Blood, Sweat, and Tears) used classifier systems with different creative variations designed to improve their performance. Much had been learned from the prior tournament, so it was not surprising that all of the new programs outperformed Rigid and most surpassed the performance of Power. Once again TiTa was in the middle of the pack, mainly because of its inability to exploit any weaknesses in its opponents.

The real surprise, however, was the performance of Feel, the program with simulated feelings. It was the overall winner of the competition. This program was identical to Power, but it now reacted to big wins or losses by amplifying the "hedonic" feedback to any rule that led to these outcomes. That is, the simulated hedonic tone was not a linear function of the payoff matrix, but acted like an hedonic amplifier that exaggerated the impact of large gains or losses. As a consequence, it reacted instantly to defection and was quick to exploit any weaknesses in its opponent.

So what has been learned from these tournaments? First of all, the success of any strategy depends upon the mix of other strategies that are in a population. Second, flexible programs that create new strategies for coping with different opponents appear to outperform less flexible alternatives like Rigid or even TiTa. One of the drawbacks of TiTa may be its inability to ever cooperate after a defection by its opponent. Less rigid strategies will periodically evaluate this possibility, and their "willingness to be the first to forgive" sometimes pays off by eliciting cooperation from an otherwise "stubborn" opponent. Third, rules similar to TiTa, such as an immediate defection following the defection of an opponent, can evolve and be implemented using simulated feelings. Acquiring Tit for Tat rules in this manner is more "humanlike" since defection following defection is not automatic but varies in probability according to the magnitude of the loss, and it arises only to cope with the behavior of specific opponents. The strategy Feel will simulate TiTa when playing against TiTa opponents, but unlike TiTa it is sensitive to the magnitude as well as the probability of gains and losses.

"Smart" programs have a semantic memory similar to the structure shown in Figure 5.1. They have a context-addressable memory where the current external context is specified by the last seven plays made by their opponent. When a similar situation occurs, the appropriate decision rules are recalled and selected in accordance with their expected hedonic outcomes, the fitness measure, that has been modified by experience. New decision rules are generated and evaluated by a creative "thought" process that explores random variations around successful designs and retains any new variations that have a high fitness.

One limitation of the current model is the absence of any simulation of internal context, the qualitative feeling state of an organism. Without this sensitivity, decisions are not guided by the current need as reflected in the hedonic state of the decision-maker. Implementing this feature would make the program's decisions sensitive to the magnitude of possible gains or losses relative to its personal survival needs. A program such as TiTa will follow the same strategy irrespective of need. In contrast, a "smart and sensitive" program would be more likely to take a risk under conditions of high personal need. For example, if the payoff from continuing to cooperate was insufficient for survival, a "smart and sensitive" program would defect; TiTa lacks this flexibility. As we will see later, humans do indeed adopt risky strategies when they are experiencing a negative hedonic tone.

Our feelings influence our reasoned decisions, and the more intense our feelings, the greater their impact. Consider the following hypothetical decision. You know a woman because she shops in the supermarket where you normally do your shopping. One day you find out that her husband is having an affair. Would you inform her of this situation the next time you met her in the store? Most people say no, and they explain their decision by stating that it's not their business. Now, what would you do if this woman was a friend? What about a cousin? How would you react if the woman was your sister? Again, most people find that as they move along this spectrum from stranger to friend (reciprocal altruist), to family member (kin), their decision changes. At some point feelings get stronger, and the decision gets modified. The observation that decisions change in the presence of strong feelings is often interpreted as emotions interfering with logical reasoning. We often say, for example, that a person in love makes "crazy" decisions. If it were just interference, however, there would be no specific pattern to these "logical errors." If reasoning, however, has an adaptive design, then the observed "errors" should be biased toward outcomes that benefit the reproductive success of the decision-maker. More accurately, feelings should bias our decisions in a manner that was adaptive during our long ancestral history. They may not, however, lead to decisions that are adaptive in today's world.

Consider the following "illogical" decision first discussed by

Cornell University economist Robert Frank.[2] Someone steals your watch, but a friend, who saw the incident, is willing to give evidence against the culprit in a court of law. It is almost certain that if you press charges and go to court, you will recover your $50 watch. Going to court, however, will mean the loss of a day's salary, $200. If you were a logical being, you would accept the $50 loss rather than the $150 loss. But humans are not logical. We get angry, and most of us want to "get the SOB" who stole our watch. Under these circumstances it simply feels better to retaliate. Indeed, if our ancestors had been unwilling to present a viable threat and follow through when someone stole their resources, they would not have survived as well as those who became angry and retaliated under the same conditions. What appears to be an "illogical" decision is really a decision based on feelings that are reliably evoked by circumstances that have consistently posed a threat to the survival of our genes.

To reduce human decisions, like the court decision, to money is an error. This is the strategy of economists, when they attempt to predict human behavior. Economists assume that money is the common currency that can be used to compare the value or utility of different decision outcomes. To a young child, however, money has no value until it can be used to acquire items that make the child feel good. Under those conditions the loss of money will result in negative feelings. The common currency that underlies all of our decisions, and indeed all of our transactions, is not money at all; it is hedonic tone. The degree of positive or negative hedonic tone associated with the various outcomes supplies the utility function that underlies all human decisions. If one outcome feels better than the other, then if the probabilities are equal, we will follow that strategy. For example, if our watch has sentimental value because it was a gift from a close friend, then its loss becomes unbearable and we would certainly go to court. Alternatively, if we really need our salary because of a necessary mortgage payment, then we may decide to forget about the watch and not go to court at all.

Many other factors, of course, could influence our feelings and therefore have an impact on our decision. Our feelings about the thief, the importance of our work, and the approval or disapproval of others can all generate hedonic tone that enters into the choice between alternative outcomes. As noted earlier, hedonic tone, the central core of all feelings, permits them to combine and interact with each other. In this manner feelings can provide the common currency that is required for complex multidimensional decision-making. Indeed, there appears to be no other possibility. Decisions are made inside human brains, and no common attribute, other than hedonic tone, exists inside a brain that is capable of combining the values associated with so many diverse factors. The shared hedonic pathway of the brain permits feelings to interact with each other, providing a basis for complex decisions involving many diverse factors.

Without feelings, decisions become difficult, if not impossible, to make. Consider the Wason decision task, which requires the application of a logical abstract rule. You are presented with four cards and informed that each card has a number on one side and a letter on the other side. You can see only one side of each card. Your task is to determine if a logical rule is being violated or not. The rule states, "If there is a vowel on one side of a card, then there must be an even number on the other side." Of the four cards shown below, which two must you turn over in order to be sure that this rule has been obeyed? Try it first.

A   B   7   8

If you are like me, and more than three-quarters of the population, you probably incorrectly selected the A and 8 cards. This task only requires that you check the truth of a very simple rule: "If p, then q." In the absence of any hedonic tone, however, it becomes a very difficult task for most of us. (Logically speaking, either a vowel or a nonvowel on the other side of the 8 card would not break the rule; the correct answer is A and 7.) Most of us don't really apply this logical rule at all, but we somehow "feel" that 8 (q) is one of the correct answers. In the real world there may be some merit in our erroneous choice of q. Consider the law "If it rains (p), then the ground must be wet (q)." Our error in the previous task is now equivalent to the false assumption that wet ground implies that it must have rained. (It could, of course, have been wet by a sprinkler.) This is not logical, but it is often a useful rule of thumb that we, and our ancestors, must have applied many times, often with favorable consequences. Humans appear to store the frequency of the co-occurrence of events in the world, and when one of these events occurs, they assume the other must also have occurred. It is not logically correct, but in the real world this potentially erroneous deduction is often better than no conclusion at all—the strictly logical alternative. Many "illogical" decision processes appears to be part of human nature, and we often jump to decisions because a particular outcome "feels" right.

Sometimes, however, humans appear to behave in a very logical manner. For instance, if we ask individuals to reason from the same "If p, then q" rule in a social domain, where strong feelings are involved, their reasoning skills improve dramatically. A rule for social exchange can be stated in the general form of "If you take the benefit, then you must pay the cost." Participants can then be asked to indicate which two of the following four cases must be investigated in order to be sure that this social rule has been obeyed:

Benefit accepted   Benefit not accepted   Cost paid   Cost not paid

In this case, over seventy-five percent of people make the correct selections: "benefit accepted" (p) and "cost not paid" (not-q).

Cosmides and Tooby have examined many variations of this social exchange paradigm and have concluded that humans have evolved specific cognitive adaptations for engaging in social exchange.[3] When performing these tasks, Cosmides and Tooby report, experimental participants experience a "pop out" effect; the correct answers are obvious to them when the rule violation constitutes cheating. Evolved feelings for monitoring reciprocal altruism provide an alternative mechanism for identifying potential transgressors. That is, when we are confronted with such a problem, we examine each option, and our feelings rapidly identify the potential infringement of the reciprocity rule. "Benefit accepted" when the cost has not been paid, and "Cost not paid" when a benefit has been accepted, are the options that evoke the most negative hedonic tone. These are immediately identified as the cases that require examination. Such decisions may "pop out" because the social circumstances match the production rule for a specific human emotion. In addition to "detecting cheaters," there are a number of other social domains, such as "threat detection" and "mate selection," where human reasoning (which relies on value systems provided by emotions) has also been found to be exemplary. All of these cases involve circumstances that engage the production rules for specific human emotions: anger, fear, and the like. In these different social domains, feelings appear to underlie our aptitude for reasoning.

REDEFINING RATIONALITY

Because humans make most decisions based on their evolved feelings and not on formal logic, their decisions are often considered to be irrational. Consider the complex social dilemma known as the Asian disease problem, first described by Tversky and Kahneman, two distinguished cognitive psychologists from Stanford and Princeton universities, respectively.[4] Participants in these studies were asked to imagine that the United States was preparing for the outbreak of an unusual Asian disease that was expected to kill 600 people. Two alternative programs to combat the disease had been proposed, and the exact scientific estimates of the consequences of each program have been made. "If program A is adopted," the participants are told, "200 people will be saved. If program B is adopted, there is a one-third probability that 600 people will be saved, and a two-thirds probability that no people will be saved." The participants were then asked, "Under these circumstances which plan would you choose?"

It is first important to note that the average expected value is the same for both plans, but one is a sure thing and the other is a gamble. If participants were to act on this observation and treat both outcomes as equally acceptable, then we would expect approximately 50 percent of them to choose plan A and 50 percent to select plan B. This, however, was not the case. In fact, when the problem is stated in the manner described above, 72 percent of participants select plan A and only 28 percent choose the risky strategy, plan B. Now here is the interesting observation. If we reword the instructions by stating that "If program A is adopted, 400 people will die, and if program B is adopted, there is a one-third probability that nobody will die and a two-thirds probability that 600 people will die," the participants' preference for the sure thing over the gamble is suddenly reversed. Despite the fact that the outcomes are exactly the same in both versions, participants now choose the risky gamble (78 percent), plan B, over the sure strategy, plan A (22 percent). This reversal in preference is known as "the framing effect." When the question is framed in terms of "saving lives" participants are risk-adverse, but when the decision is framed in terms of "losing lives," they switch to a risk-taking strategy.

Tversky and Kahneman concluded that humans are illogical because two different representations of the same decision-making problem should always lead to the same preferred outcome. By switching their preferences, the participants violate one of the fundamental principles of normative rationality: invariance. If humans are rational, and the probability and utility of decision outcomes are unchanged between the problems, then the wording of the problems should not influence their choice. The observation that they do change their preference appears (to Tversky and Kahneman) to indicate irrationality. This conclusion, however, is based on the assumption that the framing of the problem has no influence on a human's utility function (the value assigned based on the intensity of their feelings). If, as I have proposed, utility is a function of hedonic tone, then it could well be influenced by changing the hedonic tone of the decision-maker. That is, the saving-lives formulation may generate a positive feeling in the decision-maker, whereas the losing-lives formulation may generate a negative hedonic tone. If this is the case, then the decision outcomes are being evaluated from different starting positions in the two different sets of instructions. Under the negative framing condition, only the slim possibility of saving everyone has a large enough positive hedonic tone to overcome the negative hedonic tone and become the acceptable outcome. Humans, therefore, should choose the high risk decision outcome when they are already experiencing a negative hedonic tone.

The adaptive nature of this design can be more clearly seen by examining a simpler problem. Consider an animal that has to decide whether it should graze in a field that yields a small but consistent amount of food and a second field that yields, on average, the same amount of food but with much larger variance. That is, the second field sometimes has a lot of food and sometimes it has no food at all. If the animal is well fed or not very hungry, then it is adaptive to visit the first field. If it is starving, however, and the food in the first field is now inadequate for its survival, it is now adaptive for the animal to become risk taking and visit the second field. Under conditions of high negative hedonic tone (great hunger), it pays to become risk-taking, since only the expectation of the high-risk outcome is sufficient to generate a positive hedonic tone.

It was proposed in Chapter 5 that all feelings reflect potential increases or decreases in reproductive success. It follows, therefore, that any factors that influence reproductive success should also influence the intensity of an individual's feelings and thus affect decisions that are based on these feelings. Factors such as the age, sex, and reproductive status of the decision-maker, or the relationship between the decision maker and those influenced by the decision (stranger, kin, or kith), should all have a predictable impact on the decision process. The evolutionary psychologist X. T. Wang (at the University of South Dakota) has examined many of these factors and demonstrated their effects on the decision-making process.[5]

For most of their evolutionary history humans lived in small nomadic bands of hunters/scavengers and gatherers. Wang argues that compared with large groups, members of such small groups are more interdependent, engage in more social transactions, and experience higher degrees of reciprocal and kin altruism. As a consequence, group size should be a factor in human decision-making. Humans regard individuals in large groups as anonymous strangers and consider small groups to be composed of close friends and relatives. If this is the case, then the Asian disease decision discussed above should be sensitive to the size of the group that is in peril. More specifically, the potential loss of members of a small group should evoke more negative feelings than the loss of members belonging to a larger group, who are deemed to be strangers. To test this hypothesis, Wang systematically varied group size in the Asian disease problem. As predicted, the percent of participants choosing the risky outcome increased as the group size decreased from 6000 to 600 to 60 to 6. For small group sizes (60 or 6), the risky decision was the overwhelming choice, irrespective of whether the problem was worded in terms of lives saved or lives lost. The group size at which participants chose one outcome or the other equally often, under both framing conditions, was found to be about 120. This is an interesting observation because this group size corresponds to the approximate size of human hunter-gatherer bands. It appears that when group size drops below about 120, humans switch from an out-group to an in-group rationality. In a follow-up study, Wang specified that the six members of the smallest group were family members. Under these conditions the risky alternative becomes even more dominant. When asked to explain their decision, participants gave analytical arguments in the large-group context but referred to their emotions when dis-cussing the small-group decisions. Once again, only the slim pos-sibility of everyone surviving had a large enough expected value to be an acceptable outcome for participants who were experi-encing a strong negative hedonic tone.

If feelings truly underlie human decisions, then we can make another curious prediction. If feelings accurately monitor repro-ductive success, and reproductive success is always measured rela-tive to the success of others (fitness), then the intensity of feelings should also be relative to the circumstances of others. That is, a salary increase of $1,000 should be experienced as less pleasant when the recipient discovers that his or her colleagues received a $1,500 increase. Similarly, the cold weather in New Mexico should be considered less unpleasant when an individual learns that it is freezing in New York! Perhaps this is the basis of the common say-ing, "Misery enjoys company." If this is the case, and I believe it is, then information about the state of others should have a predictable and measurable effect on decision-making processes. Consider an individual who is deciding whether to apply for a new job that offers an additional $500 in income. If she is suddenly informed that she will receive a $1,000 raise in her current position, then her decision should be biased against applying for the new job. Yet strangely, discovering that her colleagues received a $1,500 increase could well tip the scales back in favor of resubmitting her applica-tion. Curiously, she is now acting against her own financial interest by applying for a job that pays $500 less than she is currently receiving. But if hedonic tone rather than monetary value underlies human decisions, her decision will be influenced by information that has no direct impact on her financial gain or loss.

In summary, human feelings appear to provide the important value system that underlies all human decisions. The shared element of feelings—hedonic tone—allows many different feelings to be combined and hence supply an overall assessment of the value associated with the various possible outcomes of a decision problem. When decisions involve social circumstance that evoke a specific emotion—such as guilt, anger, or fear—human feelings are pronounced, and decision processes are facilitated. Social decisions offer the greatest potential gains or losses to reproductive success, and humans appear to be experts in these domains. Moreover, because feelings are the omens of reproductive success (a high-stakes game), large gains and losses are disproportionately valued over small gains and losses. Furthermore, because gains and losses in reproductive success are always relative to the state of both the decision-maker and other competitors, all decision outcomes are also evaluated with respect to the hedonic state of the decision-maker and the state of his or her competitors. When all these factors are considered, humans are not irrational decision-makers at all. Rather, they make decisions based on their feelings, and these in turn specify what is, or was, in their best biological interest.

We all like to believe that we are logical human beings who can "weigh the facts" and arrive at conclusions without any emotional bias. In fact it is very difficult, if not impossible, for any of us to achieve this level of abstract logical reasoning. We would probably be very upset if we really lived in a society that was ruled by logical rules rather than human feelings. Consider how a "logical computer" would decide between two plans, one that saved two lives and an alternative plan that saved four. It would certainly favor the latter. We, however, might well select the former, especially if the two individuals were our family members. We don't even think of such a decision as being illogical, and most of us would condone this "irrational" choice if it were made by another human being. Indeed, in many circumstances the decisions of other people appear very illogical, mainly because we don't share their acquired hedonic values. We are constantly confronted with jury decisions that appear to make no sense, and we are amazed when very intelligent people decide to kill themselves and "go to another level" on a UFO that is identified by the presence of a comet! Philosophers, presidents, priests, and popes are all human beings with human minds, and their reasoning, like everyone else's, is less than rational within the framework of formal logic. Even "cold scientists" often get excited about their results, and they may feel disheartened when a treasured theory gets overthrown by the facts. With a lot of learning, however, we can eventually come to value the best hypothesis or the logical alternative, at least under most conditions. We never escape from our feelings, but we can learn to feel better about what works or best predicts the nature of the world around us. This may be as close as we can get to a useful, if not a "true," understanding of the world we live in.

NINE

∞

Passions and Illusions

IN THE BEGINNING THERE WAS A DARK, SILENT, senseless world, devoid of all meaning but pregnant with promise. Bathed from above by radiation from a distant star and heated from the core below, the ashes of past supernovas quivered with potential on the surface of a young planet. Driven in a relentless thermal dance, the primordial dust made and destroyed innumerable alliances, but only the stable survived. Strong bonds persisted and weak bonds broke, for survival of the stable was the law of the land. In ancient tidal pools crystals formed, grew, and shattered until one longitudinal crystal emerged from the fray: the replicator.[1] It may have been common, it may have been inevitable, or it may have been unique, but it was certainly significant. Blind, helpless, and thoughtless, this primeval replicator survived and reproduced. Life was born.[2]

Four and a half billion years later, Dolly was born. Modern replicators (genes) that had been extracted from an adult mammal were implanted into an enucleated egg, and the whole world marveled at their accumulated power. Dolly was conscious, and so were the scientists who created her. Their world was not dark, silent, and senseless; it was full of light, love, and meaning. What sorcery had these replicators found? What spell was now written in their secret code that could perform such alchemy? How could mere molecules transmute base matter into tastes and smells, sounds and sights, passions and desires? Don't look for the answers in books of magic, myth, or superstition, for they dwell instead within the intricate biological processes of our natural world.

Understanding the transformation of matter into mind requires two important insights. First, consciousness, like digestion, is a property of biological tissue, and conscious experiences do not exist anywhere—not in rocks, nor plants, nor computers—outside the brains of living animals. Second, like digestion and every other complex emergent property of living organisms, consciousness evolved. The first insight—that consciousness emerges from biological brains—is the most difficult to grasp, since it requires that we relinquish our intuitive vision of the world around us. Most of us believe that the world is full of light, colors, sounds, sweet tastes, noxious smells, ugliness, and beauty, but this is undoubtedly a grand illusion. Certainly the world is full of electromagnetic radiation, air pressure waves, and chemicals dissolved in air or water, but that nonbiological world is pitch dark, silent, tasteless, and odorless. All conscious experiences are emergent properties of biological brains, and they do not exist outside of those brains.

Dreams, drugs, sensory isolation, and hallucinations all reveal that input from the external world is really not necessary for conscious experiences to occur, and every attribute of a conscious mind can be altered by lesions, stimulation, or chemical modification of the neural circuitry. With your eyes closed, a gentle push on the side of the eyeball will mechanically stimulate the optic tract, and your conscious mind will experience a flash of light. Touch can evoke light, and a more powerful touch can elicit a brighter light! Our senses merely transform the energy/matter of the world into patterns of nerve impulses, but the emergent properties aroused by such patterns account for the quality and quantity of every conscious experience. The faculties of mind are nothing more and nothing less than the emergent properties of biological brains, and their nature depends on the nature of the evolved neural organization, not on the nature of the events in the world that activate them. Without consciousness there simply are no offensive smells or savory tastes, no pleasant sounds or glaring lights, for these are the passions and illusions of the mind.

The second insight—that consciousness has developed and been refined over time—requires an understanding of the adaptive process of evolution. In its simplest form Darwinian evolution requires only random variations in the structure of replicators and nonrandom selection of their progeny. Replicators make bodies and brains, but the functional emergent properties of bodies and brains dictate which replicators survive and contribute to the design of future generations. If a functional property, or any variation of that property, consistently increases the reproductive success of some individuals relative to others in the population, then any genes that contribute to this successful modification will increase in future generations. This theory of mind, evolutionary functionalism, asserts that natural selection, by favoring functional mental characteristics, has dictated the structure and organization of brains that are capable of exhibiting such mental characteristics. This is not a prescription for the evolution of a general-purpose computer or even a mind that can accurately represent and comprehend everything in its environment. Rather, this is a prescription for the evolution of minds that possess emergent properties that can be aroused by those aspects of the physical and social world that are important for gene survival. As a consequence, our minds have evolved an array of emergent properties whose sole function is, or was during the course of our long ancestral history, the enhancement of our biological survival.

The emergent properties of the brain are not meaningless or arbitrary conscious experiences, for they are the products of a relentless creative process. Unencumbered by the limitations of preconceived ideas but guided by past success, innovative evolution explores innumerable variations of successful designs in an apparently dogged search for solutions to a single pivotal problem: how to survive and reproduce in a complex changing environment. The search is blind and endless but effective, and we are the beneficiaries of its cumulative success. As a consequence, the attributes of the modern human mind are not merely the capricious properties

of neural networks; they are the highly evolved emergent properties that permitted our forefathers to survive and reproduce in their
unique and complex physical and social environment.

If the attributes of mind, like sensations and feelings, are mal-
leable emergent properties of brains, rather than rigid properties of
the environment, then they are free to evolve and become increasingly refined over generations. But natural selection "doesn't care"
if such emergent properties are, or are not, an accurate reflection
of the external world, or a product of any other criteria for that
matter; it only answers to the single constraint of gene survival. As
a consequence, our conscious minds have evolved into an array of
highly refined active filters that amplify and distinguish between
the biologically relevant aspects of the physical and social world.
Our vision of the world may be an illusion, but it's an adaptive
illusion.

THE ORIGINS OF MEANING

Far from being irrelevant by-products, our affects and emotions are
the most precious part of human nature. These evaluative feelings
define the rewards and deterrents necessary for learning, and they
are the basis upon which we make reasoned decisions. Forged by
the life and death of our ancestors, these invaluable feelings are now
an inherent part of our biological nature. Inscribed into the brain
of every human child, each priceless feeling is first evoked by the
very same circumstances that dictated life or death in its ancestral
environment. Over a lifetime the feelings themselves remain
unchanged, as the repertoire of events that elicit them grows to
mirror the dangers and opportunities that exist in the modern
world. These inner passions, however, are not mere reflections of
the biological consequences of environmental events; they have
evolved to exaggerate every small fluctuation in potential reproductive success. Within the brain, the life-and-death feedback that
guided our biological evolution has been replaced by the pleasant
and unpleasant feelings that steer our learning and reasoning.

Without feelings the physical and chemical happenings that
take place inside and outside the brains of living creatures have no
meaning in and of themselves; they just are. It is the emergent prop-

erties of our neural pathways, shaped by life and death of our ancestors, that add meaning to mere existence. Our innate and intimate
feelings, the evolved tokens of life and death, serve as the founda-
tion of all semantic networks. New events can acquire meaning
only through their learned association with these omens of our
reproductive success. Viewed from this perspective, our passions
and illusions are neither meaningless nor arbitrary, for they alone
are responsible for illuminating the darkness and adding love, color,
and meaning to the silent void of being.

This viewpoint on the human mind is markedly different
from both our commonsense perspective and the currently popular computational theories of mind. Our commonsense perspective considers the world to be full of pungent smells, rich colors,
inspiring music, and the like, and it views the contents of our mind
as almost exact replicas of that environment. In contrast, many cognitive scientists view the contents of the mind as symbolic representations of the external world that can be manipulated by
computational algorithms. Both positions are erroneous. Even if
our mental world of symbols was totally isomorphic with our
external reality, such a mind would be devoid of all meaning
because the external world is devoid of meaning. The human mind
does not acquire meaning from experiencing a host of nonsensical
environmental happenings; rather, it imposes meaning on the affairs
of an otherwise incomprehensible world. Our conscious experiences dictate the nature of our perceived reality, and our physical
and social world is inevitably interpreted, colored, and evaluated
from this evolved perspective. Our very nature destines us to experience fear and love and anger and pain in response to otherwise
"meaningless" physical and social events, and try as we will, we cannot escape from or evade such interpretations of our environment.
Indeed, to lose such evaluative feelings would be to lose the source
of all meaning and return once again to the conditions that pre-
vailed at the genesis of life.

One way to comprehend the importance of feelings is to
imagine a world without them. Imagine that you had no feelings
at all: no pain or pleasure, no love or hate, no passions or desires.
Better still, imagine you had to choose between an endless life
without feelings, or an average life span with your normal feelings.

Which would you choose? The no-feeling option is fraught with difficulties, and you probably wouldn't exist very long at all. Without feelings decisions such as whether to kill yourself become impossible. If one outcome feels just as good as any other, or more accurately if no hedonic tone is associated with either outcome, then a decision is just the toss of a coin. You might believe that you could make a "logical decision" to keep on living based on some abstract cognitive rule such as "Life is better than death." But why would you want to make a "logical decision" based on this criterion, if it feels no better than making an "illogical decision"? Of course you might "want to want" to make a logical decision, but then why would you ever want that? Either you are forced into an infinite regression of "wants," or at some point a decision must be made automatically. Indeed, animals that really don't have feelings, such as the amoebas we discussed earlier, continue to exist only because they possess a host of automatic mechanisms that control their food intake, reproduction, excretion, and every other vital function required for their survival. Life without feelings would be the life of an automatic robot, a life without meaning, not too far removed from the existence of your desktop computer!

MOVING BEYOND FREE WILL

But if our feelings dictate when we learn and how we reason about the world around us, is this not a prescription for another kind of robotic life? Do we have free will, or is our behavior determined by our nature and nurture? Before I discuss the issue of free will, it is important to clarify the difference between the terms *genetic determinism, environmental determinism,* and *biological determinism.* Genetic determinism means that all structures and behaviors are a product of our genetic makeup, and that the differences among us are a consequence of variations in our genotypes. This is clearly wrong: Identical twins have exactly the same genotype, yet although they are similar in many ways, they are also unequivocally different.

Environmental determinism is an equally extreme—and erroneous—position. It is usually not applied to the *structure* of organisms, since the difference between, say, elephants and humans obviously has a great deal to do with genetics. With respect to *behavior*, however, an environmental determinist would state that each individual's behavior is determined by his or her unique life experiences, and that the differences among individuals are strictly a consequence of variations in their environments. Some people erroneously believe that environmental determinism gives individuals more "freedom" than genetic determinism, but this is not the case. Someone controlled by the happenstance of their environmental experiences would have no more freedom that someone controlled by their genes.

Indeed, given a choice between these two alternatives, genetic determinism has more intuitive appeal, since the controlling factors, in this case, are part of our own biological nature. Environmental determinism implies that you are controlled by other individuals or other environmental factors over which you have no control. Everything you say, or do, or want has been designed into you by environmental events, particularly the actions of other people, and nothing in environmental determinism gives you the freedom to do otherwise. Like genetic determinism, environmental determinism is clearly wrong. To take an extreme example, imagine trying to teach a dog and a human how to read by giving each of them exactly the same teaching experiences. Do you really think that genetic differences are not important and that they would both learn equally well? Do you really believe that everyone would think, feel, and behave exactly the same if they had the same environmental experiences? Of course they wouldn't.

Biological determinism, by contrast, states that all behaviors and structures are determined by both genes and environmental experiences. This perspective attributes differences among individuals to differences in genes, differences in environmental experiences, or differences in the interactions between genes and environmental experiences. This position has many attractive features that are consistent with experimental findings. Genes certainly play a role in determining the structure of the brain, and learning involves additional changes to brain structure. Both processes require protein synthesis, so we should expect them to interact with each other as they jointly change brain structure and regulate the behavior of organisms. Furthermore, biological determinism does not imply that environmental factors can modify all behaviors to the same degree; this is also in keeping with experimental results. Some behaviors are readily modified by experience, whereas others are resistant to change. Finally, the potential interaction between these two factors is so great that it could account for the observed differences between individuals, and this enormous complexity could explain why precise predictions of behavior are so difficult. Nevertheless, despite its appeal, biological determinism is still determinism. Does this mean that we have no free will?

The answer to this question depends on how *free will* is defined. Most of us consider free will as "doing what we want to do." If I want to scream, and no one prevents me from screaming, then I am exercising my free will. If I live in a free society, and I am not forced to perform actions against my desires, then I have free will. In contrast, if my arm suddenly shoots up into the air for no apparent reason, then this would not be considered an act of free will, since I didn't want that event to happen. This common concept of free will also implies the ability to "change my mind." I may decide to do something and then change my mind and do something entirely different. In this case I am exercising my free will. If these statements define the meaning of free will, then it is perfectly compatible with biological determinism. This common concept of free will is a deterministic idea, because it states that "what we do" is being determined by "what we want to do." Biological determinism just provides the explanation: that what we want to do at any given time is a function of our genetic makeup and our past experiences.

At this point in the argument, an advocate of free will might object that what they really mean by free will is the ability to do something different from what they initially wanted to do—to spontaneously change their mind. Again, however, this ability is also perfectly compatible with biological determinism. If someone changes their mind because they now want to do something different from what they originally wanted to do, then they are still being determined by what they want. If, however, someone's mind changed for no apparent reason, then this would not be biological determinism; it would also not be free will, since it happened without the individual wanting it to happen. This common concept of free will is no more than dressed-up, warmed-up biological determinism.

Interestingly, the human mind is more empowered than any of these deterministic positions suggest. We are a creative species. Creativity is an inherent part of the evolutionary process and of the selection-based learning and reasoning mechanisms outlined earlier. In all three situations a random element explores variations of stored hypotheses and retains any creative solutions found to be successful.[3] The success of these procedures lies in their ability to conserve past gains and at the same time generate new creative hypotheses centered on this stored knowledge. All three adaptive processes depend upon the nondeterministic generation of innovative hypotheses and the subsequent evaluation of outcomes. As a consequence evolution, learning, and reasoning can all be viewed as creative problem-solving procedures that are inherently nondeterministic in nature.

Working together, they allow living creatures to adapt and refine their functional interactions with an enormous variety of environmental events that exhibit inherently different rates of change. Evolutionary adaptations exploit the long-term stabilities in the environment. Learning permits adaptive modifications to more rapid but consistent change, and reasoning provides a creative mechanism for adapting to future but predictable variations in the environment. For *Homo sapiens sapiens,* the source of their double wisdom depends upon making reasoned decisions, based upon learned environmental relationships, that can anticipate environmental outcomes before they occur. Such creative mechanisms endow the human mind with the nondeterministic ability to discover highly original solutions to the many survival problems encountered in a complex changing world.

Creative learning is common, but major innovation is both rare and extremely important. In our daily lives we learn and reason by creating and examining relatively minor variations around stored solutions. This procedure is adequate for discovering answers to most simple problems, because we can generate and test a large number of possibilities in a very short period of time. But occasionally major innovations arise in the minds of men and women, and these creative acts can change the quality of life for all of us. The discovery of a cure for polio, or AIDS, or the structure of DNA, or the polymerase chain reaction are all major innovations that undoubtedly have influenced, or will influence, the course of human evolution. Many individuals who may otherwise have perished are now able to survive and reproduce as a consequence of human creativity. Their offspring may then contribute to future innovations. Creativity enhances the quality of our art, music, and dance, and it is the source of inspired solutions to such major problems as transportation, communication, education, and a host of other critical issues. Each highly creative act alters our physical or social environment, and these changes redefine the requirements necessary for biological survival. In this sense we are truly the creators of our own destiny. All of this is possible because our human brain uses its highly developed emotional value system to generate a multitude of creative hypotheses based on stored information and then to appraise their potential.

Does this imply that creativity is a random process? Certainly not. The use of a random element does not imply that the outcome of the process is random. For example, it is a common misconception that evolution is a random process because it makes use of random mutations and crossovers. This is certainly not the case. The random elements in the evolutionary process generate variations around genes that have already been deemed successful by prior generations, and each new variation is then rigorously evaluated by natural selection. In a similar manner each creative new hypothesis is a variation of stored information that has been accumulated over the lifetime of the creative individual, and each one is subsequently judged by a value system that has been shaped by his or her unique experiences. Innovative learning and reasoning processes are far from random, for they are invariably steered by the past success and the evaluative feelings of their creator. The viable alternative to determinism is not free will; it is the "creativism" that arises from our selection-based mechanisms for learning and reasoning.

ENTERING THE DAWN OF FULL CONSCIOUSNESS

Those of us who view human nature through the prism of evolution are concerned with uncovering the origins of human design ("the way we are") and not with moral judgments ("the way we ought to be"). The only interest most evolutionary psychologists have in the notion of "the way we ought to be" is in explaining why some of us believe we ought to be one way, while others believe we ought to be otherwise. If feelings are the omens of our reproductive success, then we should expect them to vary according to our sex, age, reproductive status, and so on. As a consequence there is every reason to believe that human beings will not always agree on the way things "ought to be." Each individual appears to see the world from the perspective of his or her own biological platform, and it should be possible to examine and understand this relationship without defining one viewpoint as "right" and another as "wrong." Perhaps this understanding of the origins of our differences will eventually lead to a more thoughtful discussion of such controversial issues as capital punishment, euthanasia, and abortion. Indeed, the increasing popularity of democracies may be an implicit recognition of the fact that people don't agree on these and many other important issues, and a democracy may be the best vehicle for arriving at some kind of social consensus.

Most of us have no immediate awareness of the relationship between our feelings and our future reproductive success—we don't know why we feel. As a consequence we explain our behaviors by referring to their proximate cause. We might say, for example, that we acted in some manner "because we loved someone" or "because we were angry." In this respect we are all "semiconscious" creatures. We remain "semiconscious" because knowledge of the ultimate cause of our feelings is not necessary for our proximate feelings to be causally effective. We can and do make decisions based on our feelings without any awareness whatsoever of the origins of these feelings. Nevertheless, our ignorance of ultimate causation does have its consequences.

We are probably the only creatures who know that we are going to die, and our fear of death certainly influences our reasoned decisions. If we have to decide between two alternative worldviews, a natural science view of life and a supernatural alternative that offers eternal life, it is almost certain that the latter will be the more appealing alternative based on our proximate feelings. In addition, since natural science regards theories as "working hypotheses that have not yet been refuted," whereas the alternatives are often presented as "truths," it is not surprising that supernatural accounts are still the most prevalent views of the origins of human nature and the meaning of life. Indeed, such views are so ubiquitous that it is very unlikely that any individual could currently be elected president of the United States if he or she did not express some belief in the supernatural. Individuals who harbor evolutionary or other nonsupernatural beliefs are commonly assumed to be deficient in ethical values, and if such beliefs ever became prevalent, the moral fabric of society would be expected to collapse. I expect that beliefs in the supernatural will persist until human beings see themselves as moral animals, fully understand the innovative process of evolution, and come to appreciate the beauty and creative potential of their own minds.[4] We survived the Copernican revolution, and few of us now wish that we had remained within the suffocating confines of an earth-centered universe. I suspect that we will survive the Darwinian revolution and eventually come to treasure the richness of our feelings and the enormous potential for creativity that has been bestowed upon us by our ancestral history.

Human feelings are the bridge between twentieth-century humans and our primal ancestors, who hunted, scavenged, and gathered food on the African plains more than a million years ago. Like screams and gentle whispers from the past, the fears and hopes that fill our mind are first evoked by the very same circumstances that dictated life or death on the ancient African savanna. We owe our very existence to these ancestral words of wisdom that have been handed down to each of us through our long unbroken line of successfully reproducing forefathers. Now, like the ghostly oracles of ancient prophets, each fleeting conscious feeling excites or forewarns of reproductive consequences yet to come. From the ecstasy of orgasms to the inner groans of fear, these prophecies arouse the mind to act on outcomes not yet here. Not knowing where the words come from but heeding their advice, we creatively learn and reason about the world around us. Guided only by our passions and illusions, woven into the very fabric of our brain, we have sought to explore the outer reaches of our universe. Now we stand at the dawn of a new era as our feeling mind looks inward and seeks to understand itself.

ENDNOTES

∾

ONE: THE GRAND ILLUSION

1. See J. Searle, *The Rediscovery of the Mind* (Cambridge, Mass.: MIT Press, 1992) for a thoughtful discussion of the problems that confront computational theories of mind.

2. The most convincing demonstration that faculties of mind and brain are locked together comes from a series of studies on patients whose major neural connection between the brain hemispheres, the corpus callosum, has been severed. Experiments with these patients led Roger Sperry, a pioneer in split-brain research, to conclude that such subjects have two separate spheres of conscious awareness that run in parallel inside the same head. Each of these separate minds has its own sensations, memories, cognitive processes, learning experiences, and so on. In such a patient, if the word *key ring* is flashed on a screen so that *key* falls to the left of the patient's fixation point and *ring* falls to the right, then information in the left visual field (*key*) will go to the right hemisphere, and the right visual field (*ring*) will be projected to the left hemisphere. That is, the isolated "right brain" sees the word *key,* while the isolated "left brain" sees the word *ring.*

If the patient is now asked to retrieve what he or she saw from an array of objects that are concealed from sight, then the right hand, controlled by the left hemisphere, will select a ring, while the left hand will select a key. If we split the brain, we split the mind, and neither *mind-left* nor *mind-right* is aware of the other's existence. Although recent experiments suggest that the capabilities of the two hemispheres are somewhat different, each hemisphere is doubtless capable of possessing its own independent conscious experiences. See M. S. Gazaniga, *Nature's Mind: The Biological Roots of Thinking, Emotions, Sexuality, Language, and Intelligence* (New York: Basic Books, 1992) for a discussion of differences between the hemispheres.

3. The seventeenth-century philosopher René Descartes divided reality into two different kinds of substances: physical and mental. His perspective on mind, called dualism, viewed the body and brain as made of ordinary matter/energy, but considered the mind to be a nonmaterial entity composed of a completely different substance. Dualism has since become so deeply entrenched into Western religions that, outside the scientific community, it is still the most widely accepted theory of mind. Most scientists, however, have abandoned dualism because of its inability to account for the interaction between mental and physical substances—how mental events can influence the body and vice versa. If the mind is nonmaterial, then where is it? Can something that is nonmaterial be said to occupy a location in space? How can a nonmaterial mind interact with the physical matter/energy stuff of the nervous system—what would the interface be made of? If the mind doesn't interact with the matter/energy stuff of the nervous system, then it is a nonfunctional epiphenomenon, so why do we need to consider it at all? If material theories of mind are troubled by the hard problem, then nonmaterial theories of mind are plagued by a host of even harder problems.

4. A Scottish word that is a combination of *satisfying* and *sufficing.*

5. Indeed, we use a whopping one hundred pounds of ATP each day of our lives. The reason why we don't lose weight is that ADP can be recycled and reconverted back into ATP.

6. Our senses pick out particular frequencies (corresponding to red, blue, and green) and treat them quite differently from other frequencies, when in fact no defining characteristics within the electromagnetic spectrum make such points unique. Nothing in our conscious experience of redness indicates that, in the electromagnetic spectrum, it is closer to green than blue. These frequencies are arbitrary points snatched from a perfectly linear physical domain, and the relationship between them is lost in the translation. We perceive the colors violet and red as being similar, although they actually occupy opposite ends of the visible spectrum. Our conscious experiences of colors clearly don't provide an accurate linear representation of the physical world.

7. A night hunter like a snake doesn't have to be concerned about incident radiation; it has a different problem to solve. When survival depends on the ability to detect warm-blooded creatures without the benefit of reflected radiation, it is necessary to evolve sensors, and presumably subjective experiences, for detecting the infrared portion of the spectrum. Perhaps a snake "sees" a tasty glow surrounding its next meal.

8. For a discussion of the adaptive nature of color vision, see R. N. Shepard, "The Perceptual Organization of Colors: An Adaptation to Regularities of the Terrestrial World," in J. H. Barlow et al., eds., *The Adapted Mind: Evolutionary Psychology and the Generation of Culture* (New York: Oxford University Press, 1992).

9. See S. J. Gould, *The Panda's Thumb: More Reflections in Natural History* (New York: W. W. Norton & Co., 1980) for a discussion of evolutionary artifacts.

TWO: THE MOTHER OF ALL CODES

1. For a discussion of the nuptial flight, see P. P. Larson and M. W. Larson, *Lives of Social Insects* (New York: Thomas Y. Crowell, 1975).

2. A mutation is random if, for example, the factors that caused the emission of a cosmic ray that produces it are independent of the factors that caused the gene to be in the pathway of that ray. Randomness, as it is used in this context and elsewhere throughout this book, refers to events that result from the intersection of two or more independent causal chains.

3. Maternal hormones, or chemicals introduced into the maternal bloodstream, can even influence the phenotypic sex of a developing fetus.

THREE: SEARCHING FACESPACE

1. Although there is still some doubt as to whether prosopagnosia is a highly specific deficit or a mild form of a more general visual agnosia, there is no doubt that from a very early age the human brain possesses a remarkable aptitude for facial recognition.

2. J. H. Holland, *Adaptation in Natural and Artificial Systems* (Ann Arbor: University of Michigan Press, 1975). Genetic algorithms are flexible search procedures because the genotypes—the binary strings—can represent num-

bers, musical notes, facial features, or other parameters in a complex system. Using overall performance of the system as a fitness function, the algorithm will evolve the optimal, or near-optimal, values for each of the variable system components. One of the major benefits of this procedure is its ability to modify the parameters of systems that may encounter new or unexpected circumstances. For example, when the system is confronted with continual changes in environmental temperature, the settings of temperature control valves may require continuous adjustment to maintain optimal performance. Genetic algorithms are ideal for such applications because they can act as adaptive regulatory systems.

3. R. Dawkins, *The Blind Watchmaker* (New York: W. W. Norton & Co., 1986).

4. The metalevel genetic algorithm begins by generating thirty random binary strings that can each be decoded into two numerical values that specify a mutation and a crossover rate. The first of these genotypes provides the mutation and crossover parameters used by FacePrints in order to evolve a target face. In this case the target is a specific face that is selected at random by the computer. As FacePrints runs, the fitness of each generated face is computed automatically by measuring its distance from the known random target. That is, no witness is required, since the target is known and its distance from any other face in face-space can be computed mathematically. The program continues until the target face has been evolved. By repeating this process using different targets, it is then possible to determine the average number of generations required to evolve a face using the mutation and crossover values supplied by the first string of the metalevel genetic algorithm. This number serves as a measure of the efficiency in evolving a target and is used to specify the fitness of the first metalevel genotype that provided the mutation and crossover values. The entire process is then repeated using the second metalevel string to provide a new set of values for the mutation and crossover parameters, and so on. Eventually a measure of fitness is obtained for each member of the first generation of metalevel strings. The metalevel genetic algorithm then selects and breeds these genotypes in proportion to their fitness, thus creating a second generation of metalevel genotypes. Once again each member of this new generation of strings is evaluated by finding how quickly FacePrints can evolve a random face using the values of the parameters coded on its genotype. When completed, the metalevel genetic algorithm selects and breeds yet another generation of parameters. Over many generations the metalevel strings eventually converge on the optimal mutation and crossover rates for use by FacePrints. This is a convenient method for finding the best values of the mutation and crossover parameters for an efficient version of FacePrints.

5. For an evaluation of FacePrints, see V. S. Johnston and C. Caldwell, "Tracking a Criminal Suspect Through Face Space with a Genetic Algorithm," in T. Bäck et al., eds., *Handbook of Evolutionary Computation* (Oxford: Oxford University Press and Institute of Physics Publishing Ltd., 1997). Also see V. S. Johnston, "Method and Apparatus for Generating Composites of Human Faces," U.S. Patent no. 5,375,195.

6. For a discussion of selection-based learning, see H. Plotkin, *Darwin Machines and the Nature of Knowledge* (Cambridge, Mass.: Harvard University Press, 1994) and G. Cziko, *Without Miracles: Universal Selection Theory and the Second Darwinian Revolution* (Cambridge, Mass.: MIT Press, 1995).

7. This is the Church-Turing thesis.

8. J. Searle, *The Rediscovery of the Mind* (Cambridge, Mass.: MIT Press, 1992).

FOUR: RUSSIAN DOLLS

1. For a plausible neural model of this process, see N. Humphries, *A History of the Mind: Evolution and the Birth of Consciousness* (New York: Simon and Schuster, 1992).

2. The reader should be aware that Sniffers' feelings are merely a simulation. The model works only because a conscious being (the author) has assigned a meaning to the numbers being manipulated by the computer program. Indeed, without such an assignment of meaning by a conscious being, the Sniffer program, and every other computer program for that matter, computes nothing at all. Complex biological organisms possess real feelings, not simulated feelings, so meaning is an inherent attribute of their experiences. See also endnote 3 in Chapter 5.

FIVE: THE OMENS OF FITNESS

1. For an excellent review of other theories of emotion, see R. Plutchik, *The Psychology and Biology of Emotion* (New York: HarperCollins College Publishers, 1994).

2. Cziko uses the cruise control analogy to explain perceptual control theory. See G. Cziko, *Without Miracles: Universal Selection Theory and the Second Darwinian Revolution* (Cambridge, Mass.: MIT Press, 1995). In the current context, however, the analogy is being used to illustrate how high-level motor

goals, such as maintaining the speed of an automobile or performing a behavioral act, can be achieved through the use of hierarchical control systems.

3. Without feelings—affects and emotions—the world around us is a meaningless conglomerate of energy/matter, and it cannot provide the source of meaning required for a computational theory of mind. Indeed, this is the central problem with all computational theories of mind that treat the human brain as a general-purpose computer. A computer does not compute anything unless a human operator assigns a meaning to the different states of the system; meaning is not an inherent property of algorithms. Computers simply change states according to programmed rules. Unless someone assigns a meaning to these symbolic states, then no computation is possible.

Who or what assigns a meaning to brain states? In his computational theory of mind, Pinker has proposed that the "solution" to this problem lies in the "causal" and "inferential" roles of symbols. He argues that symbols can stand for things when "the unique pattern of symbol manipulations triggered by the first symbol mirrors the unique pattern of relationships between the referent of the first symbol and the referents of the triggered symbol." Of course, by this inferential role argument, a symbol cannot acquire meaning by triggering other symbols if the latter have no meaning in the first place. It doesn't matter whether an internal pattern of symbols is isomorphic with the world or not, if neither the patterns nor the symbols have any inherent meaning.

Pinker uses the "causal" argument to provide meaning for the meaningless internal symbols. This argument states that "a symbol is connected to its referent in the world by our sense organs." Here he implies that the meaning of an internal symbol can be supplied by its relationship with the external world. The problem, however, is that meaning cannot be acquired by association with the external world because there is no inherent importance in the events in the external world and no inherent purpose or intention in the laws of physics or chemistry that govern the behavior of such events. When a rock falls, there is no desire, intent, or purpose in its behavior, and the event itself possesses no inherent value or importance. So even if humans possessed a symbolic mental world that was an exact mirror image of the external world, it would still be meaningless because the external world is without meaning. Meaning can arise only through the evolutionary process, whereby conscious emergent properties of the brain—affects and emotions—come to reflect the importance of physical or social events that have consistently enhanced, or posed a threat to, biological survival. Such events have inherent value.

SIX: THE PATHWAYS OF PASSION

1. From P. D. MacLean, *The Triune Brain in Evolution: Role of Paleocerebral Functions* (New York: Plenum Press, 1990).

2. Ibid., Figure 19.1.

3. See J. LeDoux, *The Emotional Brain: The Mysterious Underpinnings of Emotional Life* (New York: Simon and Schuster, 1996).

4. See F. Crick, *The Astonishing Hypothesis* (New York: Charles Scribner's Sons, 1994).

5. See A. R. Damasio, *Descartes' Error: Emotion, Reason, and the Human Brain* (New York: G. P. Putnam's Sons, 1994).

6. S. A. Greenfield, *Journey to the Centers of the Mind: Toward a Science of Consciousness* (New York: Freeman, 1995). The quotation appears on page 137.

7. At this point the reader may wonder how resonant neural circuits can give rise to an inner subjective experience like a beautiful sunset. This question, I believe, may be based on a false premise. The function of the brain is not to recreate attributes that already exist in the world outside us. Subjective qualities like redness or beauty do not exist outside us; they exist only as emergent properties that arise from the activity within neural gestalts. Furthermore, "redness" and "beauty" are not attributes caused by resonant circuits with characteristic modes of oscillation; they are resonant activities. In exactly the same sense "temperature," as measured by a thermometer, is not *caused by* the average kinetic energy of molecules but *is* the average kinetic energy of molecules. The subjective experience and the neural activity are two different ways of viewing the same thing. If such modes of oscillation were experimentally generated within a human brain, then the owner of this brain would invariably report a subjective experience of "redness" or "beauty." This is no more mysterious than declaring that an electric current is not the *result* of electrons flowing through a copper wire but *is* the flow of electrons through a copper wire. Generating such a flow of electrons would inevitably produce an electric current.

SEVEN: MIRROR, MIRROR, ON THE WALL . . .

1. For an excellent discussion of the mental adaptations of men and women, see D. Symons, *The Evolution of Human Sexuality* (New York: Oxford University Press, 1979).

2. See D. Buss, *The Evolution of Desire: Strategies of Human Mating* (New York: Basic Books, 1994).

3. See V. S. Johnston and M. Franklin, "Is Beauty in the Eye of the Beholder?" *Ethology and Sociobiology* 14:3 (1993): 183–99.

4. See D. Singh, "Adaptive Significance of Female Physical Attractiveness: Role of Waist-to-hip Ratio," *Journal of Personality and Social Psychology* 65 (1993): 293–307.

5. The facts of sexual selection are not in question, but the origin of such preferences is open to several interpretations. Some theorists believe that a sexually selected characteristic need have no inherent value, and that even an arbitrary preference could have positive reproductive consequences. That is, if by chance alone, fifty-one percent of peahens exhibited a preference for longer-tailed males, then any female that had this preference would have an increased probability of producing both male offspring with longer tails, and female offspring who would possess her preference for longer tails—those conditions alone could lead to the evolution of longer and longer tails over generations. What started as a whim has become a widespread fashion, because wearers and admirers reinforce each other's preferences. Computer simulations have shown this to be a plausible mechanism.

There is a problem, however, with fashions that have no inherent value in themselves: They can be replaced. Consider a second female whim, perhaps for longer necks. If fifty-one percent of peahens exhibited a preference for this characteristic, then it also would enjoy a reproductive advantage, using exactly the same mechanism that led to longer tails. Now peacocks with longer necks and longer tails would soon be in more demand than their single-featured competitors. There is no reason to stop at two characteristics. If female whims alone can drive such preferences, then a host of features could evolve into stable fashions. At some point, however, such features would be in conflict with each other, for it would be impossible for a poor male to meet all of these demands simultaneously. Which characteristics would survive under these conditions? If any one of them had *real* biological value, such as parasite resistance, then it would inevitably become the dominant preference.

This thought experiment casts doubt on the origin of sexually selected traits. From the very beginning we should expect that any whim that indicates something of real value to the opposite sex (like genes that increase offspring survival) would be selected over whims that have no inherent value. Experimental evidence continues to grow in favor of this "good genes" interpretation of sexual selection. For a discussion of different viewpoints, see

H. Cronin, *The Ant and the Peacock: Altruism and Sexual Selection from Darwin to Today* (Cambridge: Cambridge University Press, 1991).

6. See A. P. Møller and R. Thornhill, *Bilateral Symmetry and Sexual Selection: A Meta-Analysis* (in press).

EIGHT: LEGACY OR LOGIC?

1. A. R. Damasio, *Descartes' Error: Emotion, Reason, and the Human Brain* (New York: G. P. Putnam's Sons, 1994).

2. R. H. Frank, *Passions Within Reason: The Strategic Role of the Emotions* (New York: W. W. Norton & Co., 1988).

3. For a discussion of cognitive algorithms for social exchange, see L. Cosmides and J. Tooby, "Cognitive Adaptations for Social Exchange," in J. H. Barlow, et al., eds., *The Adapted Mind: Evolutionary Psychology and the Generation of Culture* (New York: Oxford University Press, 1992).

4. A. Tversky and D. Kahneman, "The Framing of Decisions and the Psychology of Choice," *Science* 211 (1981): 453–58.

5. X. T. Wang, and V. S. Johnston, "Perceived Social Context and Risk Preference: A Re-examination of Framing Effects in a Life-Death Decision Problem," *Journal of Behavioral Decision Making* 8 (1995): 279–93.

NINE: PASSIONS AND ILLUSIONS

1. For a discussion of the attributes of early replicators, see R. Dawkins, *The Selfish Gene* (New York: Oxford University Press, 1976).

2. For a discussion of early life-forms, see L. Margulis and D. Sagan, *Microcosmos: Four Billion Years of Evolution from Our Microbial Ancestors* (New York: Summit Books, 1986).

3. See endnote 2, Chapter 2, for a definition of randomness.

4. Moral adaptations in humans are discussed in R. Wright, *The Moral Animal: Evolutionary Psychology and Everyday Life* (New York: Vintage Books, 1994).

INDEX